PROCESS TECHNOLOGY
SYSTEMS

Written by: Thomas D. Felder

Edited by: E. Lamar Garrett

Pearson Custom Publishing

Copyright © 2001 by Pearson Custom Publishing.
All rights reserved.

Permission in writing must be obtained from the publisher before any part of this work may be reproduced or transmitted in any form or by any means, electronic or mechanical, including photocopying and recording, or by any information storage or retrieval systems.

Printed in the United States of America

10 9 8 7 6 5 4 3 2 1

This manuscript was supplied camera-ready by the authors.

Please visit our web site at pearsoncustom.com

ISBN 0–536–62429–1

BA 992866

PEARSON CUSTOM PUBLISHING
75 Arlington Street, Suite 300, Boston, MA 02116
A Pearson Education Company

PREFACE

The Process Operator is intensively trained in the use of Standard Operating Procedures to establish and sustain Standard Operating Conditions in processes that use proven process technology. He has the responsibility to operate carefully designed facilities that safely produce high value-in-use products from a wide range of chemical feed stocks.

The typical process operator position also includes the responsibility to work effectively with a team of peers during startups, shutdowns, and process upsets.

Traditional training for the process operator position at entry level begins on the job and continues until he is qualified to fill a specific job assignment. This is normally a place on a shift team.

Initial training continues for a number of months. While the training regimen for a fully qualified process operator may well cover several years. In reality, an effective process operator-training program never ends. Process technology will surely be adapted in unpredictable ways to produce new and novel products and to improve the manufacture methods of existing ones. As a result, the process operator will train in new procedures and master new applications of process technology throughout his career in order to meet the demands of the job.

There is a growing interest in formalizing the technical qualifications for the entry-level candidate for process operator that compliments his traditional operational skills. An entry-level process operator who has a working knowledge of the chemical and physical principles at work in chemical plants is an attractive candidate for employment. He brings to the job the tools needed to detect, analyze, and communicate

accurately and precisely process conditions through his knowledge of process technology, and he uses technical terminology correctly. Effective team communications with the technical expertise that is built into the operations team is only a computer workstation away.

An Operations Team with all members educated in process technology will surely be one standard used to define the chemical industry work force of the future. Two sound concepts support the strategy to specify a technical education or the equivalent as a qualification for entry-level process operator. The process operator is ideally positioned to continuously detect, analyze, document, and communicate both measured and observed process conditions. The quality of the technical content of his observations has a significant effect on plant performance. Also, the time and cost to educate the process operator in the chemical and physical principles that he will put to immediate use on-the-job is a practical goal with a high return in performance. These are sound concepts since large chemical plants are staffed with technical specialists who rely upon the information initiated by the process operator. A team with precise and accurate data supplied by the process operator benefits the performance of both the process operator and the specialist.

The Gulf Coast Process Technology Alliance (GCPTA) has defined an educational program in process technology aimed specifically at students who aspire to be process operators. Community Colleges, with financial and technical encouragement from industry, have responded to the GCPTA guidelines by the introduction of a two-year Associate Degree Program in Process Technology. The process technology courses in the program are soundly backed by requisites in math, chemistry, and physics.

Early on, a void in suitable materials to teach process technology as defined in the GCPTA guidelines became apparent. The perfectly sound texts that are used in undergraduate engineering programs are focused on the education of future designers and builders who require knowledge of non-ideal chemical and physical behavior. The process operator needs a technical education that is centered around the well-defined process technology at work in existing facilities. The measurements and observations of process conditions in the process operator milieu depict real behavior under real conditions.

Handbooks, training films, and service manuals exist which contain useful procedures and information, but the content is directed primarily to the care and routine operation of specific equipment combinations.

The basic process technology offered in this text is selected to fill the need of students who aspire to be the technically qualified process operator candidate of the future.

In summary, the foremost objective in teaching process technology is a presentation of the principles of physics and chemistry that are integrated into a presentation of the unit operations that are commonly used in most chemical plants. A corollary objective is to instill the use of technical symbols and terms that accurately and concisely describe process conditions and the process variables used to control those conditions in safe and profitable plants. Adequate teaching materials is a start.

TABLE OF CONTENTS

CHAPTER 1	INTRODUCTION TO PROCESS TECHNOLOGY	1
CHAPTER 2	HEAT AS A SOURCE OF ENERGY	43
CHAPTER 3	COMBUSTION AND FURNACES	115
CHAPTER 4	STEAM GENERATION	149
CHAPTER 5	STEAM DISTRIBUTION AND CONDENSATE RECOVERY	171
CHAPTER 6	REFRIGERATION	201
CHAPTER 7	DISTILLATION	233
CHAPTER 8	WATER TREATMENT	301
CHAPTER 9	COOLING TOWER	359
CHAPTER 10	EXTRACTION DECANTATION	433
CHAPTER 11	CRYSTALLIZATION	453
CHAPTER 12	CHEMICAL REACTIONS	463

CHAPTER 1

INTRODUCTION TO PROCESS TECHNOLOGY

The Process Technology used in petrochemical plants exploits well-defined principles of physics and chemistry. Plant resources are organized to establish and maintain a safe and productive workplace.

The Process Operator runs his area through consistent use of standard procedures to sustain standard conditions. The process variables of pressure, temperature, flow and a number of others are only visible evidence of the physical and chemical changes that are carried out inside the equipment where raw materials are being converted into high value-in-use products.

It follows that a working knowledge of the basic technology in use in plant processes is paramount to an understanding of how well a plant is performing. Knowledge of process technology also provides an important insight into probable cause when a process goes astray.

This chapter is designed to introduce the student to several aspects of process technology that the successful student has to manage. The first involves the symbols and terms that are used to communicate process data effectively. The second aspect is the consistent need to quantify information in precise and accurate terms. If one does not use the technical language correctly, he cannot communicate properly with those who do. Unfamiliar terms and symbols often give the unfair impression that the subject matter is complex. The student will discover in this chapter that he is aware of, and even uses a number of elements of process technology in his daily life. The challenge in the study of

process technology is to see how basic chemical and physical principles are applied to do useful work. A good place to start is with heat energy.

HEAT ENERGY

All matter has the ability to absorb and de-absorb heat energy. The heat energy always flows from a higher temperature (heat source) toward a lower temperature (heat sink). The amount of heat energy in matter can be quantified by the temperature difference (ΔT) and the thermal properties of the material in question (See Figure 1).

The heat energy that matter absorbs is in two basic forms. The first, called sensible heat, is measured by temperature. The temperature rises and falls with the addition and removal of sensible heat. There is no change in phase

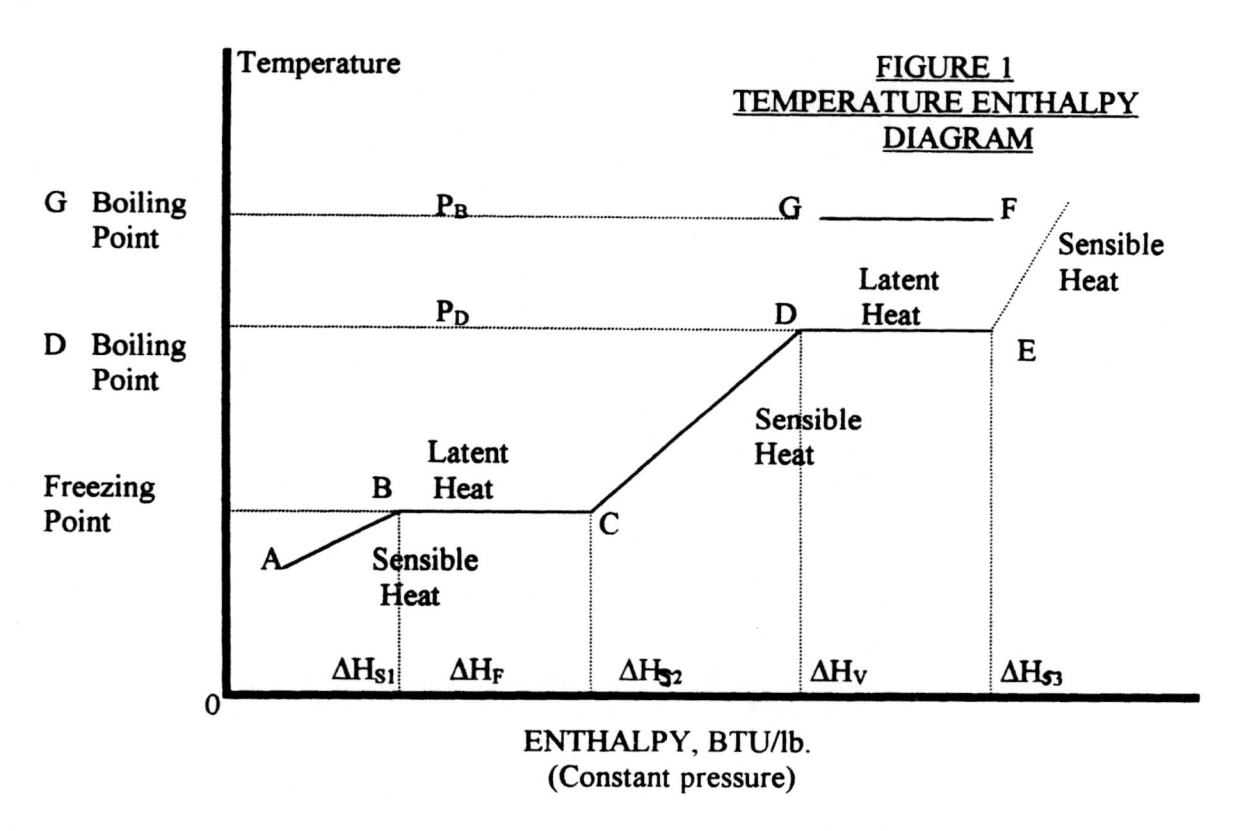

**FIGURE 1
TEMPERATURE ENTHALPY DIAGRAM**

Sensible heat is transferred within and between solids, liquids, and gases where there is a source and a receiver identified by a temperature difference. The movement and spread of heat energy is due to thermal mixing called diffusion.

The second form of heat energy is called latent heat and always involves a phase change. The quantity of latent heat is large per unit weight and is always absorbed at constant temperature as long as the pressure is constant.

The enthalpy temperature diagram in Figure 1 can be used to explain and show the relationship of several important terms and symbols that define heat energy.

The temperature and enthalpy vary with each material under study. A general plot is used here for study purposes. Lets start with a substance in its solid state at Point A. As heat energy is added, the temperature of the solid increases and will start to melt at Point B, which is called the freezing point.

The total sensible heat (ΔH_{S1}) absorbed by the solid can be calculated by the equation:

$$\Delta H_{S1} = M\ C_p\ \Delta T$$

    ΔH_{S1} is the increase in heat content of the solid as sensible heat, BTU

    M is the mass or weight being heated, lbs.

    Cp is the heat capacity of the solid material being heated, BUT/lb./°F

    ΔT is the temperature increase between points A and B.

Heat capacity, Cp is the critical term used in the equation and is unique to every material. It reads, as the amount of heat energy required to raise one pound of material one degree in temperature.

As additional heat is added starting at point B, all the solid is converted to liquid at point C. Two important points here. The conversion from solid to liquid occurs at constant temperature as the material absorbs latent heat. The change in state is called a

phase change and the latent heat input is called the heat of fusion. The quantity of heat transferred at constant temperature as the phase change occurs can be stated as:

$\Delta H_F = M\, H_F$

    ΔH_F is the total heat input, BTU

    M is the mass, lbs.

    H_F is the heat of fusion, BTU/lb.

At point C, the solid has been converted to liquid and the total heat content has increased by the sensible heat put into the solid plus the heat of fusion.

As additional heat energy is absorbed starting at point C, the temperature increases to point D. The heat energy absorbed is sensible heat. The temperature at point D is called the boiling point.

The sensible heat required to increase the temperature of the liquid from B to C is expressed by the equation:

$\Delta H_{S2} = M\, C_P\, \Delta T$

    ΔH_{S2} is the total sensible heat, BTU

    M is the weight being heated, lbs.

    C_P is the heat capacity of the liquid, BUT/lb./°F

    ΔT is the temperature difference.

As additional heat is added at point D, the liquid vaporizes until all the liquid is converted to vapor at point E. The conversion of liquid to vapor is also phase change at constant temperature and pressure and the latent heat input is called the heat of vaporization, H_V. The quantity of heat is calculated by the equation:

$$\Delta H = M H_V$$

ΔH is the total heat energy transferred

M is the weight vaporized, lbs.

H_V is the heat of vaporization, BTU/lb.

In summary, a solid at point A has been heated to a vapor state at point E, which included two sensible and two latent heat transfers (phase changes). What makes this physical process so important is that the temperatures and quantities of sensible and latent heat are reproducible and reversible for every material. If heat is removed from the substance starting at point E, the quantities of sensible and latent heat will backtrack in step with a reduction in temperatures to point A. The heat of condensation is equal to the heat of vaporization. The heat capacity (C_P) used to calculate sensible heat is the same value when heat is added or removed so the quantities of sensible heat de-absorbed are identical. The total heat energy removed to go from vapor at the boiling point to solid at the freezing point is the same as the amount of heat energy added going from solid to vapor.

If sensible heat energy is added beyond the boiling point, the temperature continues to rise and the vapor is said to be <u>superheated</u>.

There is another very useful aspect of heat energy that occurs at the boiling point. The boiling point of a liquid increases as the external pressure on the liquid is increased. Liquid that vaporizes at boiling point D in Figure 1 will vaporize at boiling point G if the pressure is increased from P_D to P_G.

Another useful fact is the behavior of solutions. Solids dissolved in liquids tend to elevate the boiling point and suppress the freezing point.

MORE ON BOILING POINT AND FREEZING POINT

Boiling point is the temperature where a liquid vaporizes with the input of latent heat. The boiling point temperature is unique to each pure substance and to mixtures of known concentrations. The boiling point temperature increases and decreases with pressure but the heat energy necessary to vaporize the liquid at different pressure levels is only marginally effected. What is important in many plant operations is that the heat of vaporization is exactly reversed as the heat of condensation.

The freezing point is the temperature where a liquid converts to a solid, which is described as a phase change. Every pure substance has a distinctive freezing point. The latent heat removed is called the heat of fusion. Pressure does not change the freezing point but freezing liquids can exert enormous pressures in piping systems.

SUBLIMATION

There are substances like carbon dioxide that change phase directly from the solid state to the vapor state, when heat energy is absorbed. Materials are said to sublime when undergoing change from solid to vapor.

HEAT OF FORMATION

Another form of heat energy is released or consumed when a compound is formed from the elements. This energy is aptly named the heat of formation. The heat of formation has been measured and documented for a great many chemicals. This form of heat energy is not particularly relevant to the responsibilities of process operator except for one important use. When two or more chemicals react to form different chemicals, the heat energy consumed or released is the algebraic difference between the heats of formation of the products and the heats of formation of the reactants.

The process operator does have the responsibility to control the heat energy transferred in chemical reactions.

There are other forms of heat energy, usually small, that may be important in specific systems. Gases cool when expanded and heat up when compressed. Solutions may heat up when the solute is added to the solvent. As individual unit operations are studied, all forms of heat energy involved will be explained.

THE SYSTEM CONCEPT

A system is simply any combination of operations within defined boundaries. The system concept is what makes the use of energy and material balances to define physical and chemical changes a practical tool in analyzing plants operations.

A pump can be a system. A tank can be a system. The pump and tank can be combined into another, larger system. A process has a defined work area that is a system. The process operator sees each operation under his control as a system, which is a valuable tool in comparing what is to what should be.

MASS AND ENERGY TRANSFER

All unit operations are dependent upon mass and energy transfer. Mass transfer is a term used to describe the processes that separate mixtures, effect a change in phase or state, and a necessary step in any change is chemical composition.

Energy transfer for most of the operations in a petrochemical plant is the movement of heat energy into, or out of mass and includes the heat of reaction. The operations studied in this course use mass and energy transfer to do useful work.

There are three conditions needed to transfer mass and energy. They can be expressed in the broadest of terms as:

Driving Force

Contact

Time

The Process Operator who has a working knowledge of one or more of those three elements present in his equipment is in an excellent position to sense change. Observe what change is being made compared to the change expected.

MATERIAL AND ENERGY BALANCE

The Process Operator uses the principles of heat (energy) and material (mass) balances to control and follow the performance of industrial processes. His primary tools are operating procedures to set and control instrument readings that display the conditions under which materials and energy are being put to use.

Natural laws of chemistry and physics are inflexible on two technical points within the conditions that all industrial processes operate.

1. Energy cannot be created or destroyed but can be changed in form. Steam energy used to generate electricity is an example.
2. Mass (materials) cannot be created or destroyed but can be changed in composition. The chemical reaction is an example.

When these two natural laws are applied to a closed system, all mass and energy that enters that system must leave that system or accumulate within that system.

The major points can be shown on a diagram.

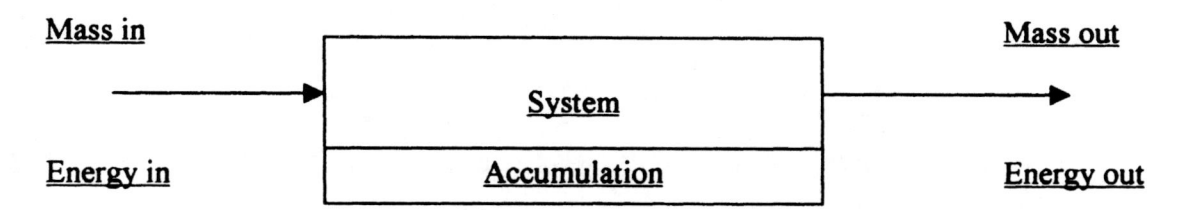

Mass in →	System	→ Mass out
Energy in	Accumulation	Energy out

Energy in = energy out + accumulation

Mass in = mass out + accumulation

In continuous processes, the accumulation is usually zero. Therefore,

Mass in = mass out

Energy in = energy out

ENERGY BALANCE

Energy is somewhat more difficult to account for than mass. However, the basic principle is the same. All the energy put into a system must come out or accumulate. Accumulation is rarely desirable and is a sign of a potential process upset. This is common occurrence where the energy is contained in material that does not react properly.

What makes the energy balance so important is the immense quantity of energy being used in petrochemical plant. The potential for destructive forces is high if the input and removal of energy are not controlled. Any unexpected change in temperature is a process variable the process operator must account for.

The simplest exchange is sensible heat measured by an increase in temperature. There is no latent heat involved and no accumulation.

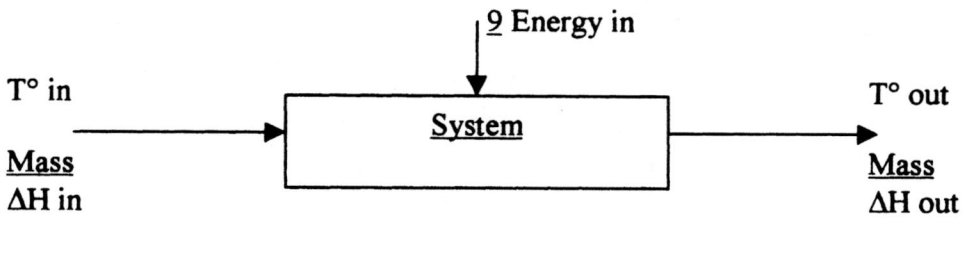

```
                        Q Energy in
                             ↓
  T° in              ┌──────────────┐           T° out
  ─────────────────▶ │    System    │ ─────────────────▶
  Mass               └──────────────┘           Mass
  ΔH in                                         ΔH out
```

$\Delta H = (M)(C_P)(\Delta T)$

ΔH = change in enthalpy, BTU/HR. (H out – H in)

C_P = heat capacity, BTU\lb.\°F

ΔT = T° out - T° in   °F

M = mass flow lb.\HR.

Q = heat added BTU\HR.

It is not important where the heat input comes from. The balance is:

$Q = \Delta H$ BTU\HR.

This is the principle of heat transfer in shell and tube heaters and coolers.

Another example of an energy balance that is encountered frequently involves both sensible and latent heat exchange. No accumulation. Consider an evaporation process.

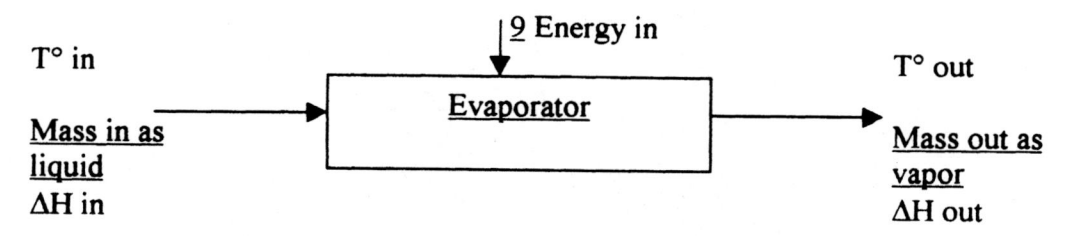

```
                          Q Energy in
                              ↓
   T° in                ┌───────────┐            T° out
   Mass in as      ──▶  │ Evaporator│  ──▶       Mass out as
   liquid               └───────────┘            vapor
   ΔH in                                          ΔH out
```

The liquid stream entering at T° in, is vaporized and exits at T° out. Assume that liquid is present during the vaporization and the phase change takes place at T° out. Then the sensible heat is all used to heat the liquid to the boiling point T° out and super heat is not involved. The total heat in Q is the sum of the sensible heat and latent heat.

$$\Delta H_T = \Delta H_S + H_V = Q$$

ΔH_S = sensible heat, BTU\HR.

$\Delta H_T = (M)(C_P)(\Delta T) + (M)(H_V)$ ΔH_T = Total heat exchanged BTU\HR.

M = mass in and out, lb.\HR.

C_P = heat capacity of the liquid mass in BTU\lb.\°F

ΔT = T° out - T° in, °F

H_V = heat of vaporization BTU\lb.

Q = heat input, BTU\HR.

For this example, it is not important where the heat comes from. The balance is:

$$\Delta H_T = Q = (M)(C_P)(\Delta T) + (M)(H_V)$$

Sensible heat + Latent heat

We will study this example and its reverse in vaporizers, condensers, refrigeration, and cooling towers. Also, in steam generation and steam distribution.

One of the most important examples of the energy balance involves the chemical reaction. Chemical reactions will be studied in depth later when the groundwork is properly laid.

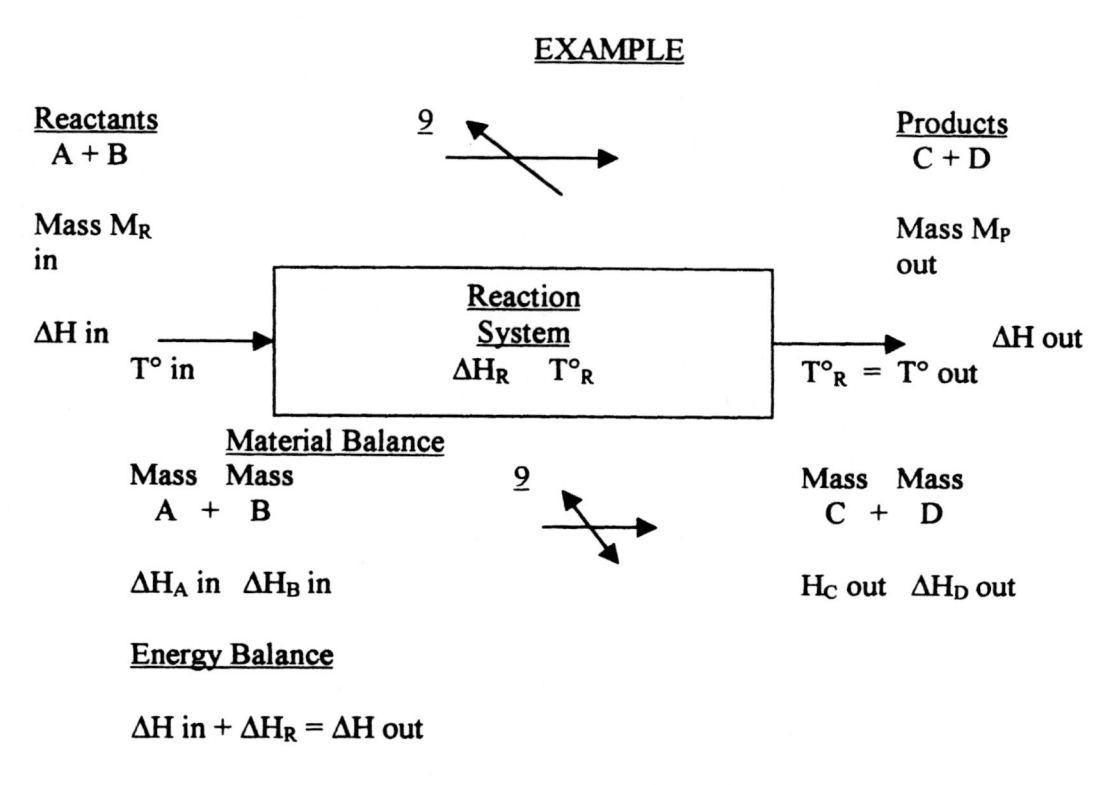

EXAMPLE

Reactants		Products
A + B		C + D
Mass M_R in	Reaction System ΔH_R $T°_R$	Mass M_P out
ΔH in		ΔH out
$T°$ in		$T°_R = T°$ out

Material Balance

Mass A + Mass B → Mass C + Mass D

ΔH_A in ΔH_B in H_C out ΔH_D out

Energy Balance

ΔH in + ΔH_R = ΔH out

ADIABATIC REACTION

DURING the adiabatic reaction, there is no exchange of heat in and out of the system. <u>Zero heat in</u> and <u>zero heat out</u>. In practical use, the adiabatic reaction is exothermic.

The reactants A + B are converted to products C + D in stoichiometric quantities. The *reaction produces heat energy H_R. The heat of reaction must be absorbed by the reactor contents since no heat energy can get in or out. The temperature increases and is controlled at steady state level and is the exit temperature $T°$ out. In summary, the sensible heat increase is equal to the heat of reaction.

This is where the use of a reference temperature is useful as a baseline. Since most reaction engineering data is reported at 77°F (25°C) it is a commonly used reference temperature.

A Summary Table may be helpful in understanding reference temperature. 77°F

Heat in, ΔH_{in} Heat out, ΔH_{out}
Reactants   Products

$\Delta H_{in} = \Delta H_A + \Delta H_B$ $\Delta H_{out} = \Delta H_C + \Delta H_D$
$\Delta H_A = (M_A)(Cp)(\Delta T)$ $\Delta H_C = (M_C)(CP)(\Delta T)$
$\Delta H_B = (M_B)(CP)(\Delta T)$ $\Delta H_D = (M_D)(CP) \Delta H_D$

$\Delta T = (T° \text{ in} - 77°)$ $\Delta T = (T° \text{ out} - 77°)$

ENERGY BALANCE

$$\Delta H_{in} + \Delta H_{out}$$

The interesting observation is that if T° in is below 77°F, the ΔH_{in} is a negative quantity i.e., it contributes to the capacity to absorb the heat of reaction.

We will not study the calculation of the heat of reaction from heats of formation at this point. It will be covered in the study of reactions.

EXAMPLE WITH HEAT REMOVAL

Assume that the reaction takes place in stoichiometric quantities, 100% conversion, and no phase change. Reference temperature is 77°F.

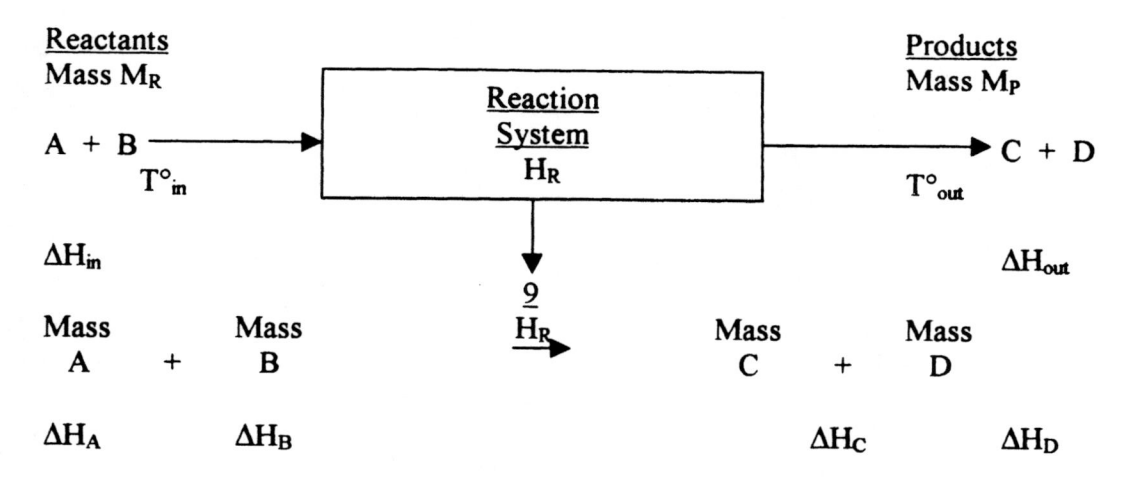

```
Reactants                                           Products
Mass M_R                                            Mass M_P
                   ┌──────────────┐
A + B  ──────────▶ │   Reaction   │ ──────────────▶ C + D
   T°in            │    System    │    T°out
                   │      H_R     │
                   └──────┬───────┘
ΔH_in                     │                         ΔH_out
                          ▼
                          Q
Mass          Mass       H_R       Mass         Mass
 A     +       B       ──────▶      C     +      D

ΔH_A          ΔH_B                 ΔH_C          ΔH_D
```

Many reactions produce more heat energy than the system can absorb. In that case, the heat must be continuously removed at a rate to maintain steady state conditions. This is conveniently done with a circulating pump and a heat exchanger.

The reactants A + B are converted to products C + D in stoichiometric quantities. The reaction produces heat energy H_R.

Summary Table

Reactants	Products
$\Delta H_{in} = \Delta H_A + \Delta H_B$	$\Delta H_{out} = \Delta H_C + \Delta H_D$
$\Delta H_A = (M_A)(C_p)(\Delta T)$	$\Delta H_C = (M_C)(C_p)(\Delta T)$
$\Delta H_B = (M_B)(C_p)(\Delta T)$	$\Delta H_D = (M_D)(C_p)\Delta H_D$
$\Delta T = (T°in - 77)$	$\Delta T = (T°out - 77)$

$\underline{Q}$ = Heat removed

Energy Balance

$$\Delta H_{in} + H_R = \Delta H_{out} + \underline{Q}$$

FIGURE X
VAPOR PRESSURE

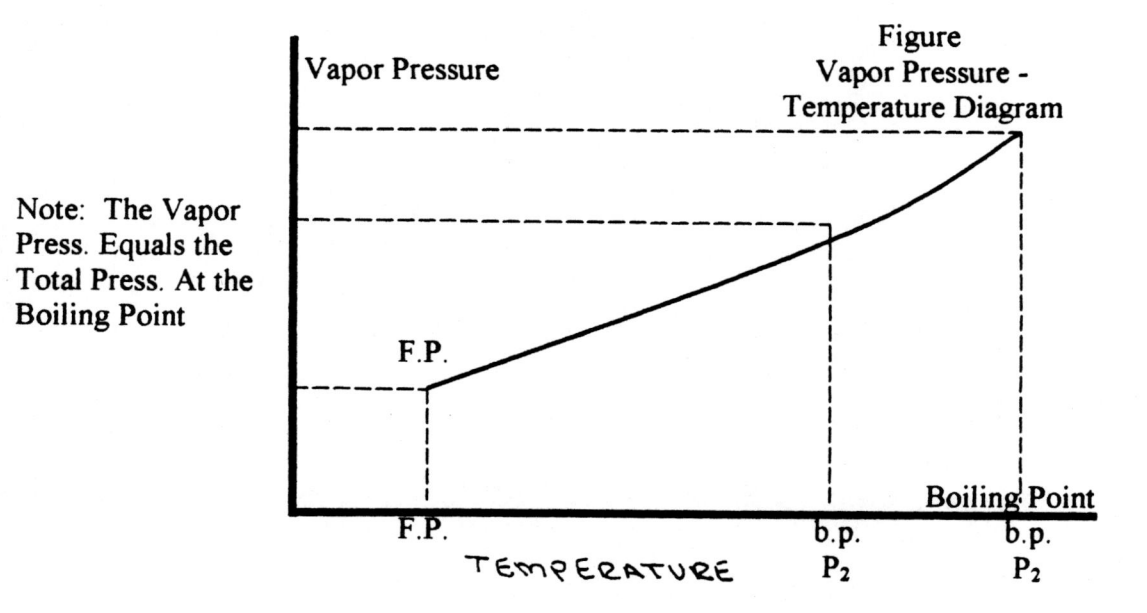

Figure: Vapor Pressure - Temperature Diagram

Note: The Vapor Press. Equals the Total Press. At the Boiling Point

(Graph: Vapor Pressure vs. Temperature, showing curve from F.P. to Boiling Point, with b.p. at P₂)

As a liquid absorbs sensible heat energy, the temperature increases from the freezing point to the boiling point (See Figure X). The molecular activity also increases. In a closed system maintained at constant pressure, energized molecules are driven from the liquid and establish a concentration and a partial pressure in the gas that is in contact with the liquid. This partial pressure is called the vapor pressure.

The vapor pressure increases with the temperature. Where the vapor pressure equals the total pressure of the gas in contact with the liquid that is called the boiling point. What makes this so useful is that the vapor pressure and the boiling point increase or decrease with an increase or decrease in the total pressure on the system. This is a physical principle used in steam generation and refrigeration.

SOLUBILITY AND MISCIBILITY

When a material is dissolved in a liquid, it is said to be soluble in that liquid. The liquid is called the solvent and the material dissolved is the solute. The mixture is called a solution. When the solvent contains all the solid it can hold, the solution is said to be saturated.

The solubility of solids and liquids in liquids increases with temperature. Pressure has little or no effect (See Figures BELOW). The solubility of gases in liquids decreases with temperature but increases with pressure.

Solubility is an important factor in water treatment, extraction, and waste disposal.

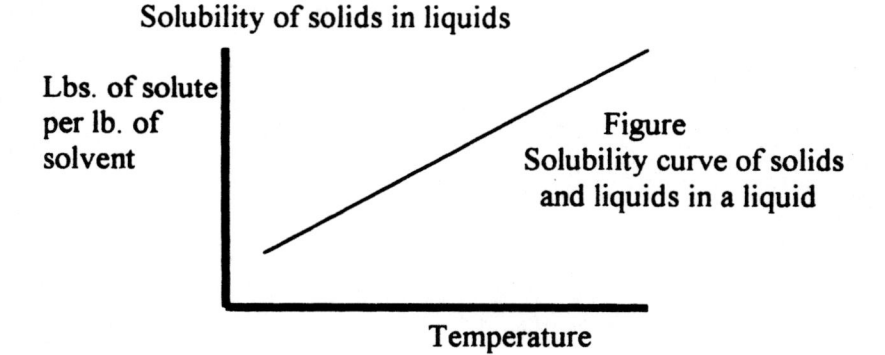

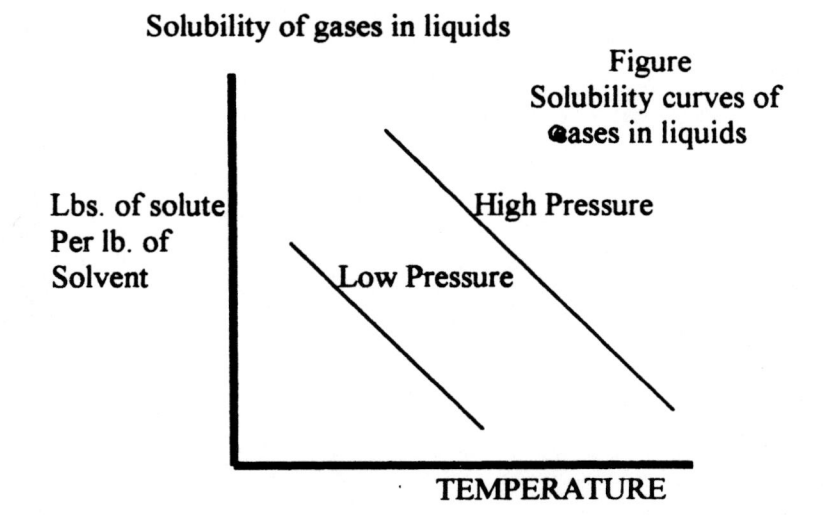

Solubility of solids in liquids

Figure
Solubility curve of solids and liquids in a liquid

(Y-axis: Lbs. of solute per lb. of solvent; X-axis: Temperature)

Solubility of gases in liquids

Figure
Solubility curves of gases in liquids

High Pressure
Low Pressure

(Y-axis: Lbs. of solute Per lb. of Solvent; X-axis: TEMPERATURE)

33

MISCIBILITY

When two liquids can be mixed in all proportions and lose their individual properties to form new properties in a mixture, the liquids are said to be miscible. Vinegar and water form a typical mixture that is said to be miscible in all proportions.

When a material has little or no solubility, it is said to be insoluble.

When two liquids will not mix, in any proportion, they are said to be immiscible. The classic example is oil and water.

GAS LAWS

A series of laws cover the interaction of the variables of pressure, temperature, and volume in gases. It is impossible to change one variable without changing one or more of the sister variables. Complete control of gas properties is central to many basic operations in a petrochemical plant. To the Process Operator, an understanding of the gas laws is a key to understanding gaseous behavior in process equipment.

BOYLE'S LAW

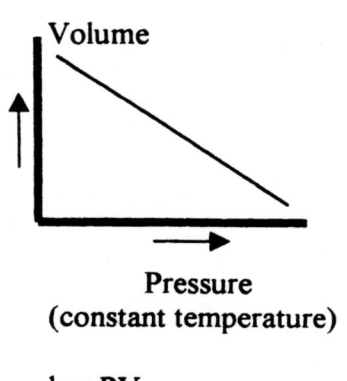

Volume

Pressure
(constant temperature)

$k = PV$

k is constant
P is the absolute pressure
V is the volume

Boyle's Law states that a change in absolute pressure at constant temperature creates an equivalent but opposite change in volume. The product of pressure times volume is a constant.

CHARLE'S LAW

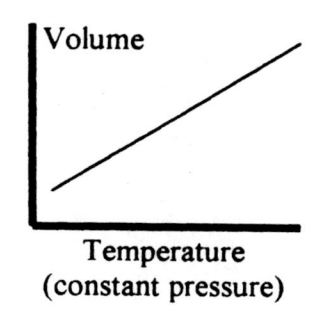

Volume vs. Temperature (constant pressure) — linear increasing graph.

$$k = V/T$$

k is constant
V is the volume
T is the absolute temperature

Charle's Law states that an increase or decrease in temperature at constant pressure creates an equivalent increase or decrease in volume.

The combination of Boyle's Law and Charle's Law creates the general gas law.

$$P_1V_1/T_1 = P_2V_2/T_2$$

DALTON'S LAW

A number of important gases exist as mixtures such as air. The typical gaseous process stream contains multiple gas components. Dalton's law states that the total pressure exerted by a mixture of gases is the sum of the partial pressures of each component in the system.

$$P_T = P_1 + P_2 + P_3 + P_4 \cdots P_N$$

Another very useful fact is that the partial pressure of each component is proportional to its molecular concentration.

AMAGAT'S LAW

The logic applied to derive Dalton's Law on additive pressures is also true for the additive property of partial volumes of components in a gas mixture.

$$V_T = V_1 + V_2 + V_3 \cdots V_N$$

The obvious conclusion that is also important is that volume fractions and molecular fractions are identical quantities in gas mixtures.

LIQUID – VAPOR LAWS

There are two laws that are useful in understanding the relationship between gases and liquids.

HENRY'S LAW

Henry's Law states that a gas dissolved in a liquid is proportional to the partial pressure of the gas in contact with that liquid.

P_O = partial pressure of gas

$P_O = k N_S$

N_S = concentration of gas in a liquid

RAOULT'S LAW

Raoult's Law states that the partial pressure of a volatile substance in the vapor above its liquid is proportional to the molecular concentration of the volatile substance in the liquid, times its vapor pressure at the temperature of the liquid.

| Partial press. of a Component in vapor | = | Molecular conc. of the component in liquid | × | Vapor pressure of component at liquid Temperature |

It can be concluded that the partial pressure is proportional to volume and molecular concentration.

CHAPTER 2

HEAT AS A SOURCE OF ENERGY

The Process Operator uses heat energy to do useful work in steam generation, refrigeration, distillation, chemical reactions, and additional unit operations. The heat exchanged in reboilers, heaters, coolers, and condensers are typical examples.

The technology used to both add and remove heat from a process stream is identical. The issue in most processes is how effectively the transfer of heat is controlled. The thermal properties that can be used to define the heat content of materials are involved in most of the processes in a chemical plant.

Two more lessons involving the use of heat that are important to the process operator in training are conservation and safe usage. Conservation of heat energy is a significant economic factor. Heat energy, not properly used, is a major contributor to hazardous conditions.

The experienced and knowledgeable process operator is in a position to take immediate action that conserves energy and prevents hazards.

THE NATURE OF HEAT

The atoms and molecules that make up matter are in continuous motion. Add heat and the motion, predictably, increases. Remove heat and the molecular motion, predictably, decreases. When heat content is increased, it is said to be absorbed. When heat content is decreased, it is given off or de-absorbed. However, two important principles must be satisfied in the movement of heat. First, there must be a donor and a receiver. Heat will not be given off that has no place to go. Secondly, heat moves only a from high temperature source to a low temperature receiver.

Thermal properties used to define the heat content of materials are involved in most of the operations in a chemical plant. The separation of components in a mixture occurs in distillation and heat does the work. A refrigerant both absorbs heat and discharges heat to do work on a process stream. Chemical reactions, steam generation, and evaporation are additional uses of heat to do work.

The work carried out in many unit operations depends upon efficient absorption and, equally important, efficient discharge of heat through the precise control of temperature change. This is possible because the energy in a closed system is fixed. Any additional heat added to a system must come out of the system or accumulate. The fact that energy exchanged must balance at the heart of process control.

Heat is the most universal source of energy found in chemical plants. There is the need to add heat to some unit operations or to remove heat from others. The important principle is that the exact amount of heat put into an operation has to come out of that operation or accumulate. The large amount of latent heat that accompanies a phase change from liquid to vapor is an important process used to move large amounts of

energy from one place for use in another place. An important principle is that the latent heat in a change of liquid to vapor and vapor to liquid is identical but moves in opposite directions. This principle is used in distillation, refrigeration, and steam generation. When there is a need to put heat into a process at one point and take out that same heat at another location, the quantities must balance.

Heat is a common factor involved in number of hazards such as burns, fires, and explosions. Facilities used to carryout chemical reactions contain the potential energy being converted to kinetic energy to do major damage if the heat or reaction is not rigorously controlled. Another particularly hazardous situation is when hydrocarbon air mixtures ARE allowed to exist within the flammability or explosive limits.

Olefins, which are heavily used to produce plastics, have the ability to decompose with a violent release of energy in the presence of an initiator such as oxygen. The most effective approach to instill an understanding of heat in and out of unit operations is to study plant application for similarities and differences. The process operator must understand the heat in use in each unit operation in order to understand the functions of the equipment and the control logic. This section includes movement of heat, how it is controlled, and how it is quantified. The principles learned here are used in the study of the role heat plays in specific processes.

UNDERSTANDING TEMPERATURE AND HEAT CONTENT

The study of heat must include a clear understanding of the use of temperature as a tool to measure and control the heat content of process streams.

Temperature is measured and read on scales that are divided into equal segments call degrees (See Figure 1).

Using water as an example, the Celsius scales (°C) is divided into one hundred degree segments between the freezing point and the boiling point of water. The Fahrenheit scale (°F) representing the same change in heat content is divided into one hundred and eighty degree segments.

FIGURE 1

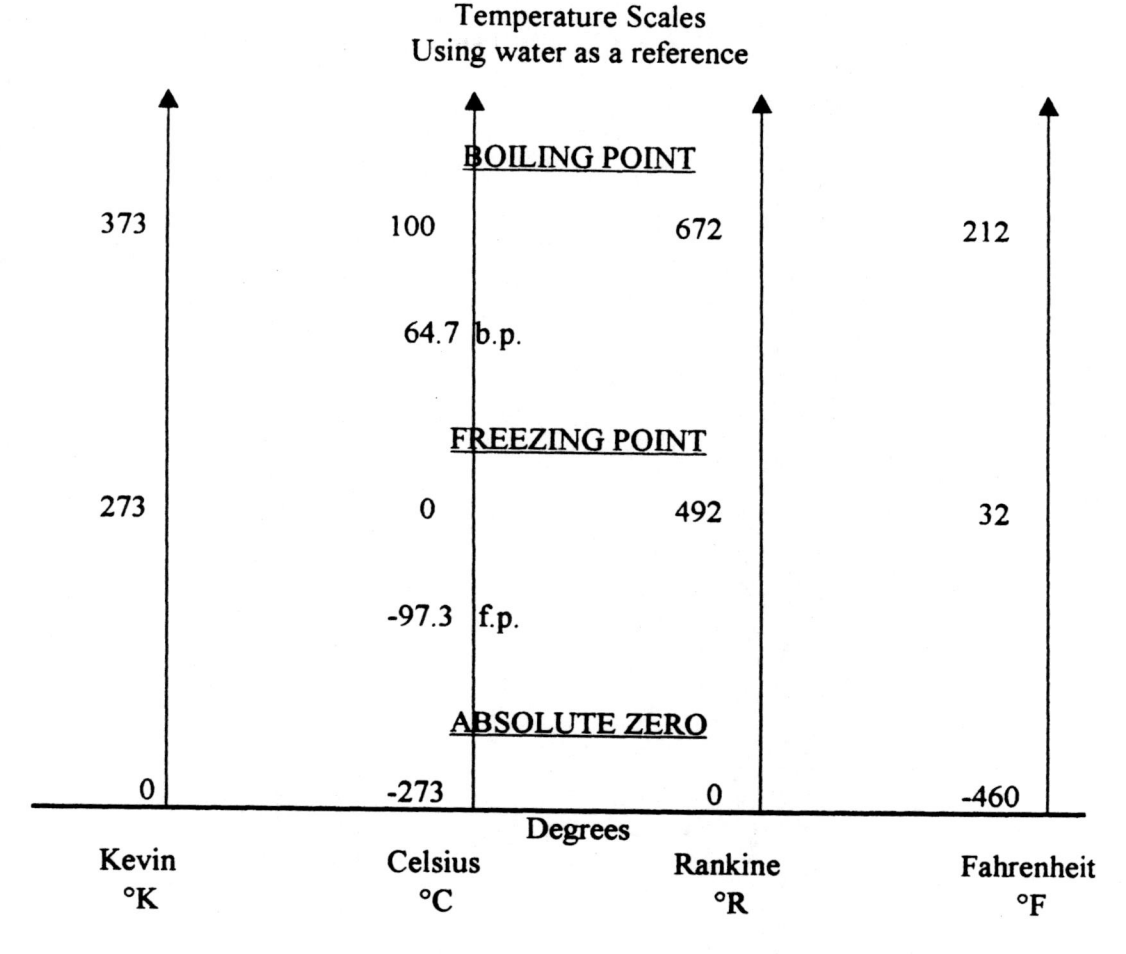

Temperature Scales
Using water as a reference

Kevin °K	Celsius °C	Rankine °R	Fahrenheit °F
373	100	672	212
	64.7 b.p.		
273	0	492	32
	-97.3 f.p.		
0	-273	0	-460

BOILING POINT (100 °C / 212 °F)
FREEZING POINT (0 °C / 32 °F)
ABSOLUTE ZERO

Degrees

Notes: In plant usage, degrees Celsius and degrees centigrade are interchangeable with degrees centigrade the more commonly used term.

An identical chart showing the freezing point and boiling point of methanol would read 64.7°C and –97.8°C. Boiling point temperature changes with the system pressure, which will be explained in the study of specific processes. The data in figure 1 are taken at atmospheric pressure, 14.7 PSIA, 0 PSIG. Both scales represent the same amount of heat being absorbed or given off between the boiling and freezing points of water. The significance of this fact is that each degree change on the Celsius and Kevin scales measures 1.8 times the change in heat content that is measured by each degree change on

the Fahrenheit and Rankine scales. A change in temperature of a material indicates a change in heat content that can be easily calculated using the British Thermal Unit (BTU).

The British Thermal Unit (BTU) is a universally fixed quantity of heat defined as the amount of energy required to raise the temperature of one pound of water one degree Fahrenheit (°F), THEREFORE the specific heat (SP. HT.) of water is expressed as one BTU/lb./°F. In comparison, 1.8 BTUs of heat would be required to raise the temperature of one pound of water one degree Celsius (°C).

The change in temperature is the only field information needed to calculate the change in the heat content of a substance when there is no change in phase or state involved. The substance remains in the original form as liquid, vapor, or solid throughout the temperature change.

The change in heat content with an increase in temperature but no phase change can be stated as:

$$\Delta H = {}^*C_p \Delta T$$

    ΔH is the increase in heat content, BTU/lb.

    *C_p is the specific heat, BTU/lb./°F

The change in heat content with a <u>decrease</u> in temperature and no phase change can be stated as:

$$-\Delta H = {}^*C_P \Delta T$$

    $-\Delta H$ is the decrease in heat content, but/lb.

    *C_p is the specific heat, BTU/lb./°F

*C_p is OFTEN called heat capacity, ΔT is the change in temperature, °F

If the temperature readings are in degrees Celsius (°C), the ΔT must be multiplied by 1.8 to convert to degrees Fahrenheit (°F).

Determination of the heat content of a substance that involves a phases change is a bit more subtle since there is no change in temperature through the phase change.

The energy absorbed or given off during a phase change is called latent heat. The absorption or discharge of latent heat occurs at constant temperature, which is unique for each substance. The transfer of latent heat occurs at two temperatures – the freezing point and the boiling point. The temperature that a phase change begins stays constant until the phase change is complete. Again using water as the example shown in Figure 1, the melting of ice occurs at the freezing point of 32°F with the addition of the latent heat of fusion (H_F). Vaporization of water occurs at the boiling point of 212°F with the addition of the latent heat of vaporization (Hv).

A phase change is reversible with a reverse in the heat content. Water vapor condenses at the boiling point of 212°F with removal of the latent heat of vaporization (H_V). Liquid water freezes at 32°F with the removal of the latent heat of fusion (H_F).

The change in the total heat content of a substance when heat is added can be expressed as:

$$\Delta H = {^*C_p} \Delta T + H_F + H_V$$

> ΔH is the change in total heat content, BTU/lb.
>
> *Cp is the specific heat, BTU/lb./°F
>
> ΔT is the change in temperature, °F
>
> H_F is the heat of fusion BTU/lb. (If no change in solid to liquid occurs, H_F = 0)

H_V is the heat of vaporization, BTU/lb. (If no change of liquid to vapor occurs, $H_V = 0$)

If temperature change (ΔT) is measured in °C, multiply ΔH by 1.8.

When heat is being removed from a system, the change in heat content can be expressed as:

$-\Delta H = *C_p \Delta T + H_F + H_V$

$-\Delta H$ is the heat removed, BTU/lb.

$*C_p$ is the specific heat, BTU/lb./°F

ΔT is the change in temperature, °F

H_F is the heat of fusion, BTU/lb. (If no freezing occurs, H_F is 0.)

H_V is the heat of vaporization, BTU/lb. (If no condensation occurs, H_V is 0.)

ΔH (heat in) and $-\Delta H$ (heat out) have the same numerical value in BTU/lb. that is unique for each material going through the same temperature and/or phase changes.

THE THREE WAYS TO TRANSFER HEAT

Several principles must be learned in order to understand the movement of heat.

- Heat moves from a higher to a lower temperature.

- For heat to transfer, there must be both a recipient to absorb heat and a donor to provide heat.

- Sensible heat is absorbed accompanied by an increase in temperature. Sensible heat is given off accompanied by a decrease in temperature.

- Latent heat is absorbed or given off during a phase change with no change in temperature.

RADIATION

The most universal transfer of heat is by <u>radiation</u>. Radiation can be described as "waves of heat energy". The sun radiates heat in waves. So does a toaster, fireplace, and an electric heater. Radiant heat can pass through a vacuum or a medium like air and be absorbed by an object. An automobile, for example, can be heated by the sun to a temperature above atmospheric. A common industrial application of radiant heat would be a furnace.

Characteristics of Radiant Heat

- Occurs at high temperatures.

- Does not need matter (solid, liquid, or gas) to transfer from one <u>body</u> to another. The sun is an example.

- The heat can be described as <u>waves</u> of energy.

- Heat is said to be <u>absorbed</u> by matter.

- The controlling mechanism in furnaces, radiators, and fireboxes.

A heated object radiates heat based on its composition, size, and temperature. There does not have to be a receptor specified for this form of energy emission. The heat impacts on all objects in the line of sight. How much radiant heat an object absorbs depends upon the intensity of the heat emission, the composition of the object, and its size. The only practical controls on emission of radiant heat are intensity of the source, exposure to the source, and the distance from the source.

Materials that do not absorb radiant heat readily are also of considerable value. Fire brick is an example of a material that reflects most of radiant heat back toward the source.

The value of radiant heat lies in the high temperature that heat transfer occurs. Chemical plants use radiant heat by combustion (controlled fire), in fireboxes, furnaces, and direct fired exchangers. The generation of steam in boilers is a common use of radiant heat in a firebox. Radiant heat is also used to crack hydrocarbons into fragments that can be rearranged into valuable intermediates or products.

CONVECTION

Heat transfer by convection is a major factor in chemical plant operations. Convection occurs at all times between fluids (gas or liquid) of different temperatures when molecular contact is made. The warmer molecules transfer heat to the cooler molecules until the entire mixture reaches an intermediate temperature.

Characteristics of Convection

- Occurs at all temperatures but is most useful to transfer heat at moderate to low temperatures.
- Transfer occurs when warm and cool materials collide. Contact is important. AND IS USUALLY PROVIDED BY MIXING.
- This is the primary heat transfer mechanisms in fluids.

The quantity of heat transferred depends upon the specific heat (Cp) of the material, the temperature difference (ΔT), frequency of contact between hot and cold molecules, and the time for heat to transfer.

The overall heat transferred can be determined by a mass and energy balance. (SEE FIGURE Y)

MASS BALANCE

$$M_A + M_B = M_M$$

ENERGY BALANCE

$$-\Delta H_A = \Delta H_B$$

$$-M_A C_{pA} \Delta T_A = M_B C_{pB} \Delta T_B$$

$$\Delta T_A = T_A - T_M \qquad \Delta T_B = T_M - T_B$$

FIGURE 4

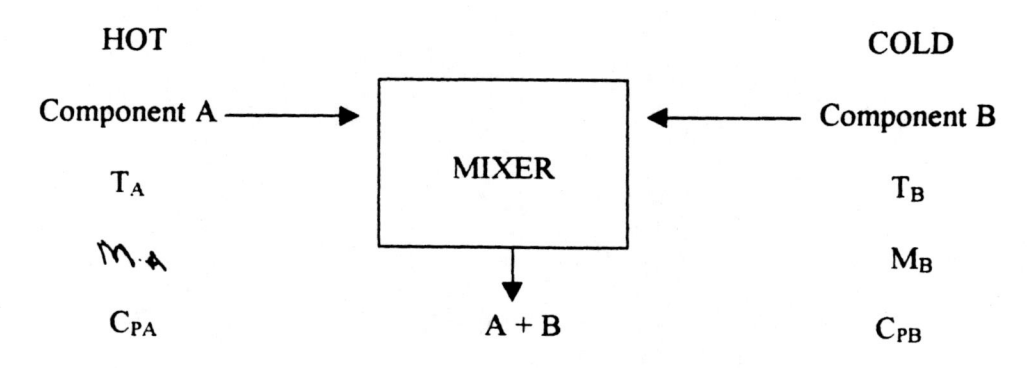

HOT		COLD
Component A →	MIXER	← Component B
T_A		T_B
M_A	↓	M_B
C_{PA}	A + B	C_{PB}

T_M is the temperature of A+ B

C_{PM} is the specific heat of A+ B

CONDUCTION

Heat transferred through a solid is called conduction.

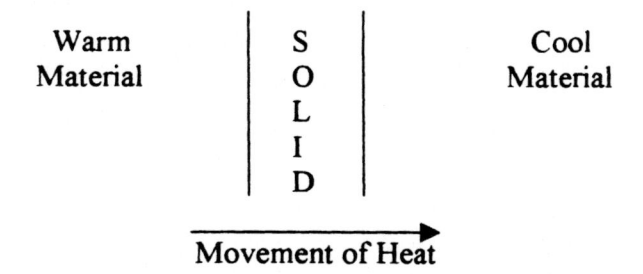

```
   Warm        | S |        Cool
  Material     | O |      Material
               | L |
               | I |
               | D |
        ──────────────▶
        Movement of Heat
```

Heat moves from the warm material through the solid to the cool material.

Separation of the heat donor and the heat receiver is the major use of conduction.

<u>Characteristics of Conduction</u>

- Conduction is the major mechanism of heat transfer through a solid.

- Conduction occurs when there is a temperature difference (ΔT) across the solid. ΔT is the driving force that moves all heat.

- Conduction is limited by a physical property of the solid called conductivity.

MOST COMMONLY USED THERMAL PROPERTIES

TERMS, SYMBOLS, AND UNITS

Specific Heat, CP, BTU/lb./°F (OFTEN REFERRED TO AS HEAT CAPACITY)

Quantity of heat required to increase the temperature of one pound of mass of a substance one degree Fahrenheit.

Sensible Heat, Associated with a temperature change, ΔH or Q_{IN} BTU/lb.

The heat absorbed that increases the temperature of a substance with no phase change involved.

Latent Heat, Associated with a phase change

A. Heat of Fusion, H_F in BTU/lb.

The heat required to convert a pound of solid substance to a pound of liquid or to convert a pound of liquid substance to a pound of solid at constant temperature.

B. Heat of Vaporization, H_V in BTU/lb.

The heat of vaporization is the energy required to change a unit weight of a liquid to a gas, or a gas to a liquid. It is stated in calories/gram or BTU/lb. It is a practical way to "store" a large amount of heat in a substance by vaporization that can be recovered when the substance is condensed. Steam is an example. Heat of condensation is the exact reversal of heat of vaporization.

Thermal Conductivity, k

Conductivity is a measure of the amount of heat that a solid material transfers through itself when a temperature difference (ΔT) exists:

$$k = \frac{BTU}{(HR)(FT^2)(°F/FT)}$$

k = coefficient of conductivity

Where
HR = Time of one hour

Ft² = Transfer Area of one square foot

°F = A degree of temperature driving force

Ft = Thickness (distance) in feet of the solid through which the heat is transferred.

K is a very important factor in the calculation of the transfer of heat through a solid.

Coefficient of Thermal Expansion (Contraction), FT/FT/°F

All unrestrained matter expands when it is heated and contracts when it is cooled. The amount a material contracts or expands can be calculated using the coefficient of thermal expansion, which reads feet of expansion or contraction per foot of original length per degree of temperature change.

If we are interested in a volume change such as in the content of storage tanks, the cubical or volumetric coefficient is used, FT³/FT³/°F. The growth or contraction reads cubic feet per cubic feet of original volume per degree change in temperature.

Equipment that is restrained or liquid filled systems without temperature for expansion can be damaged by the large forces developed as expansion or contraction occurs.

Reference Temperature

All matter contains heat at any temperature above absolute zero. Since it is not practical to reference heat down to that level, heat content of a substance is calculated based on a convenient reference temperature. For scientific work, 25°C (77°F) is commonly used as a reference. A reference temperature simplifies the heat balance calculation since the heat capacity of a substance can be documented at a reference temperature in the literature for convenience.

HEAT TRANSFER ACROSS A WALL

Heat is transferred in Industrial Processes in a variety of equipment using the principles of conduction, convection, and radiation. Heat can be transferred in the same operation by more than one principle method but one usually dominates. The objectives in plant operations are to consistently control the rate and amount of the heat being transferred.

A universally used method of heat transfer in an Industrial Process is conduction from one stream to another through a divider call a "<u>wall</u>" (See Exhibit 1). This divider can be a pipe wall, tube wall, or the wall of a vessel. Convection is also important, so the materials on each side of the wall need to be continuously well mixed to maintain a uniform temperature difference across the wall. This is necessary for control of the rate heat is being transferred.

Two important principles of heat transfer are that heat energy only moves from high temperature to low temperature. The driving force is the temperature difference across the wall.

The resistance to heat transfer through a wall consists of a number of factors. Some are difficult to measure and often change with operating conditions. The Process Operator must understand the individual resistances to heat transfer across a wall, where any or all can effect the overall heat transfer performance. The technical expertise needed to analyze overall heat transfer performance across a wall is an important troubleshooting skill.

EXHIBIT 1

The Five Resistances to Heat Transfer Across a Wall

1. <u>Medium</u> at temperature T_1

    T1 temperature is kept uniform by mixing under laminar flow conditions

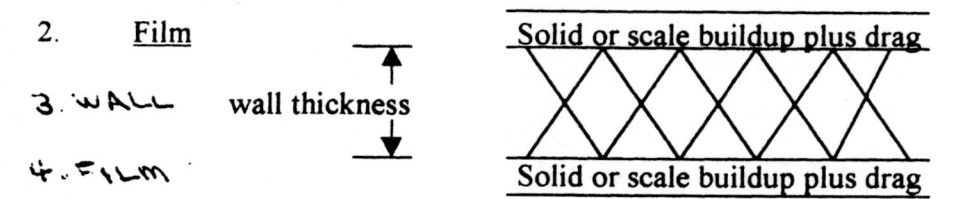

2. <u>Film</u> — Solid or scale buildup plus drag

3. WALL — wall thickness

4. FILM — Solid or scale buildup plus drag

    T_2 temperature kept uniform by mixing under laminar flow conditions.

5. <u>Medium</u> at temperature T_2

The material in the wall resists the transfer of heat. The resistance of the wall material can vary widely and is a unique thermal property of a material called conductivity.

The resistance to the transfer of heat through a material with a uniform composition is proportional to the distance it has to go. In a wall, that distance is the thickness measured in feet.

Another important factor that controls the <u>amount</u> of heat that is transferred is the surface area exposed to the high and low temperature. It has been shown, all other factors being equal, that heat is transferred uniformly over the available area if good mixing of the medium occurs on each side of the wall. Ideally, the heat transferred is directly proportional to the transfer area.

The temperature difference across the wall is the driving force that causes the heat to transfer from the hot to the cold side. Again, with good temperature uniformity on both sides of the wall, the heat transferred per degree of temperature difference (ΔT) per

square foot of surface area is constant. All of the factors that effect heat transfer through the wall itself are combined into one constant called thermal conductivity, k.

K = BTU/HR/FT./FT²/°F

>BTU/HR is the quantity of heat transferred in one hour.

>FT. is the thickness of the wall in feet.

>FT² is a surface area of one square foot

>°F is a temperature difference of one degree Fahrenheit across the wall.

There is resistance to the transfer of heat into and out of the wall by a film on the wall surfaces. The films are caused by drag through adherence of the medium on each side of the wall. Viscosity of the medium is a major factor in the amount of drag and film thickness that a material exhibits.

The film thickness is affected by the movement (velocity) of the medium along the wall. Plug flow creates a thicker film while laminar flow results in a thinner film. The thin film offers less resistance to heat transfer. A simplified way of looking at this layer of material or film is that it adds to the distance or thickness of solid material that the heat must travel through from medium to medium.

A portion of the overall film resistance to heat transfer can be attributed to the buildup of material on the wall surfaces called scale. Scale is often a poor conductor of heat and even a thin coating will have a negative effect by increasing the overall resistance to heat transfer.

THE OVERALL HEAT TRANSFER COEFFICIENT

In summary, the efficient transfer of heat through a wall is determined by a number of variables that can change over time. Measurement and control of individual variables would be impractical, if not, impossible. However, steady state control of heat transfer across a wall is the heart of the heat exchanger, AND is critical to chemical plant operation.

The answer has been found in the use of an overall coefficient stated in the following relationship (See Exhibit 2).

$$U = \frac{Q}{A \, \Delta T}$$

1. U is the overall heat transfer coefficient, BTU/Hr/Ft²/°F. It is the net result of overcoming the five resistances that control the movement of heat through a wall. From the basic equation, the value of U is determined by the amount of heat being transferred (Q), the area across which the heat is transferred (A) and the temperature driving force (ΔT).

2. Q is the heat being exchanged between media on each side of the wall. It is referred to as the heat load in BTU/Hr. The heat will always move from the higher to the lower temperature. The rate is affected by each of the five resistances but can be measured as one overall coefficient, U.

3. A is the area across which the heat must move across the wall. For a tube, the outside surface area is used. The outside area of a tube is calculated by multiplying the cross sectional area by the tube length.

$$A = \frac{(\pi D^2)(L)}{4}$$

A = Outside tube surface area, Ft²
π (pi) = constant, 3.1416
D = tube diameter, ft.

L = tube length, ft.

4. For the total surface area in a tube bundle, the surface area of one tube is multiplied by the number of tubes. ΔT (Delta T) is the temperature difference. The heat can be transferred in either direction across the wall but always moves from the higher to the lower temperature. The temperature difference is the driving force that determines the rate of heat transfer.

EXHIBIT 2

Typical Tube Wall

Overall Coefficient			
T_A ↑ ΔT U ↓ T_B	Medium B ———— Area ———— Tube Wall ——————— Medium A	Q ↕	Heat moves in the direction from high to low temperature.

Five Individual Resistances to Heat Transfer

1. Medium A
-------------------------- 2. Film
3. Tube Wall
-------------------------- 4. Film
5. Medium B

THE BASIC HEAT EXCHANGER (See Exhibit 3)

The basic heat exchanger is designed to transfer a known amount of heat that is calculated from the process flow and the temperature difference ΔT between the inlet and exit flows. The transfer area needed to move a specific amount of heat depends upon the overall resistance to heat transfer. Calculation of the individual resistances to heat transfer is a difficult task and only necessary for design purposes. In plant oprations, the overall heat transfer coefficient is monitored which works well as a measure of performance. A typical U has been determined and documented for many heat and

cooling situations. The effective area, (A) can be calculated from the basic equation using the U that best fits the process conditions.

EXHIBIT 3

Basic Heat Exchanger

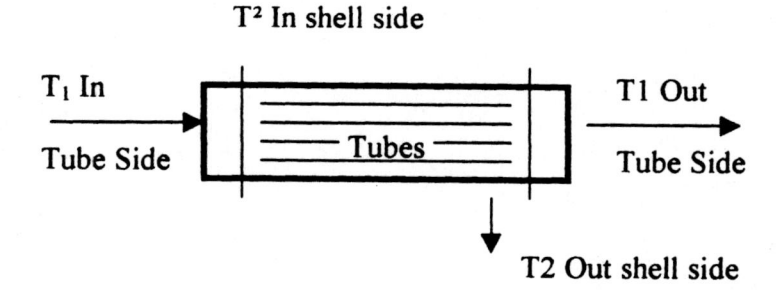

T^2 In shell side

T_1 In → Tube Side | Tubes | → T1 Out Tube Side

↓ T2 Out shell side

A = Outside surface of one tube x number of tubes.

The Process Operator monitors the overall coefficient to follow the performance of a heat exchanger. The task is simplified by a correlation with easily measured process variable, such as coolant temperature, steam pressure, and process flow (See Exhibit 4).

EXHIBIT 4

Deterioration of U at Constant Process Flow

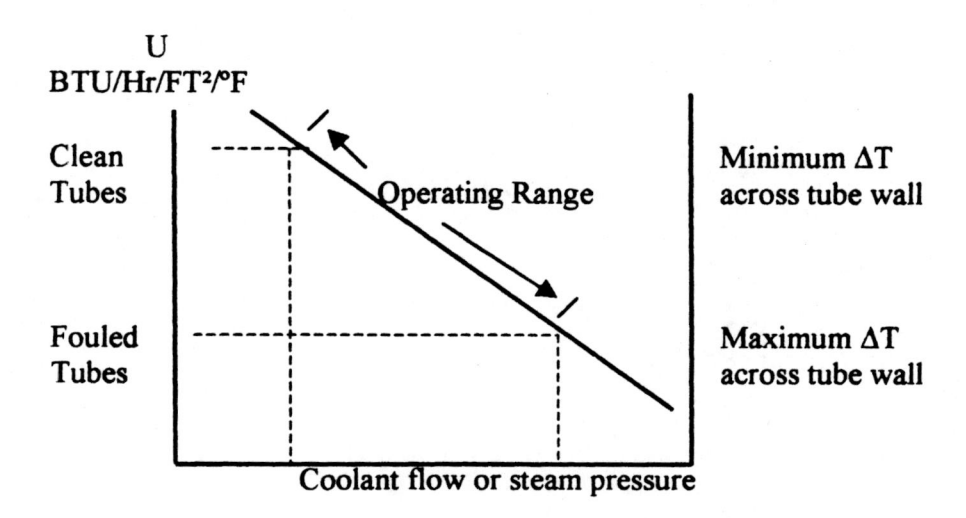

U BTU/Hr/FT²/°F

Clean Tubes — Operating Range — Minimum ΔT across tube wall

Fouled Tubes — Maximum ΔT across tube wall

Coolant flow or steam pressure

As U deteriorates, it requires more coolant flow or higher steam pressure to transfer a fixed amount of heat, (Q) for a fixed process flow, (M).

Understanding Temperature Difference

The temperature difference (ΔT) is the driving force behind the transfer of heat energy across a wall. However, the temperature difference used in charts or plots to relate to the heat being transferred does not seem to match the instrument readings. The reason is that the heat being transferred is driven by the overall average or mean temperature (See Exhibit 5).

EXHIBIT 5

Mean ΔT Across The Heat Exchanger

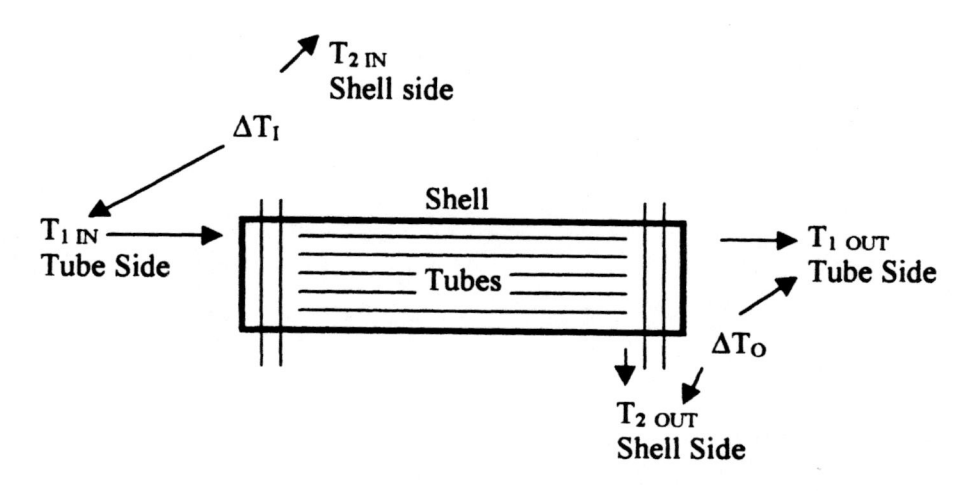

T1 IN is changing to T1 OUT and T2 IN is changing to T2 OUT. The working or mean temperature difference (ΔTM) between T1 and T2 is somewhere between the inlet and exit temperature. Mathematically, the difference has been shown to be:

$$\Delta T_m = \frac{\Delta T_I - \Delta T_O}{LN \frac{\Delta T_I}{\Delta T_O}}$$

Where
ΔT_M Mean temperature difference
ΔT_I Largest temperature difference between T_1 and T_2
ΔT_O Smallest temperature difference between T_1 and T_2
L_N Natural logrithum

T_1 and T_2 in and out area easily measured by instruments but ΔT_M is calculated as the true driving force that transfers the heat energy and must be calculated.

It is not necessary to be as exact in following U as a calculation of ΔT_M provides. A useful estimate of the temperature difference ΔT is the differences between the average temperatures in and out.

$$\Delta T = \frac{\overset{IN}{T_1 + T_2}}{2} - \frac{\overset{OUT}{T_1 - T_2}}{2}$$

As U deteriorates, it requires a higher steam pressure (condensing temperature) or a higher coolant flow to transfer a fixed amount of heat, (Q) in a fixed process flow.

Tracking U is a useful way to follow and predict heat exchanger performance. Any of the five resistances to heat transfer, or a combination could be the cause of loss in U.

However, since the wall material is constant, the most likely loss in U is due to a higher film resistance or poor convection (mixing) in one of the media.

A knowledge of heat transfer will assist the process operator in the determination of actual cause. But the real benefit of tracking is the ability to predict the useful life of the heat exchanger between remedial maintenance.

DETERMINATION OF HEAT LOAD, Q

If the value of Q is known, the efficiency of heat transfer is easily followed by monitoring temperature and flow readings.

The laws of Conservation of Energy and Mass state that mass and energy entering a system must come out or accumulate. Since there is no accumulation of mass or energy in a typical heat exchanger.

Mass in = mass out

Energy in = energy out

So to determine Q, one needs to know the amount heat lost or gain by either medium (See Exhibit 6). What one medium gains in heat, an equivalent quantity of heat is lost by the second medium.

EXHIBIT 6

Heat Lost or Gained in a Heat Exchanger

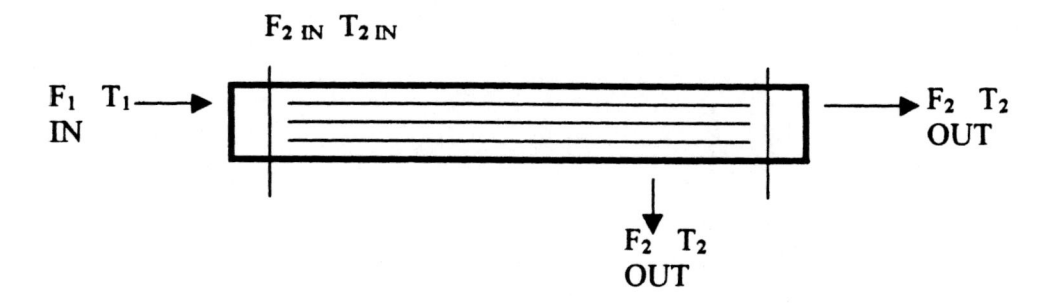

F_2 IN T_2 IN

F_1 T_1 IN → [heat exchanger] → F_2 T_2 OUT

F_2 T_2 OUT

The Basic Equations that define the heat gained or lost by the media are:

$Q_1 = (F_1)(C_P)(\Delta T_1)$

$Q_2 = (F_2)(C_P)(\Delta T_2)$

$Q_1 = Q_2$

The simplest calculation is to use the coolant temperature difference and coolant flow on a cooler and the process temperature difference and process flow on a heater.

Notes: The heat capacity (C_P) must be in the same units as the flow and temperature. Typical would be BTU/lb./°F.

ΔT is always the positive difference between T_{IN} and T_{OUT}.

Both sensible and latent heat is effectively transferred in the shell and tube heat exchanger.

BASICS OF HEAT TRANSFER EQUIPMENT FEATURING THE SHELL AND TUBE EXCHANGER
INTRODUCTION

The transfer of heat in an industrial process is most often through a solid from one fluid to another. The wall keeps the source of heat or the coolant separated from the process stream. The heater or cooler can be as simple as a coil in a tank. It can be the jacket on a pipe or vessel wall. However, the heat transfer coefficient may be low in such simple designs and temperature control may be poor due to poor mixing (convection).

The solution to achieve efficient heat transfer (best utilization of ΔT and transfer area) has been the development of a very sophisticated family of shell and tube heat exchangers that can be designed to handle a wide range of process services and heat transfer conditions. The principles of heat transfer place few restraints on size, materials of construction, and location. Process control by use of the common variables of flow, temperature, and pressure is excellent. The shell and tube is a dependable <u>workhorse</u>, and is heavily used in petrochemical plants.

This lesson will concentrate on the shell and tube exchanger.

The principles of heat transfer studied up to this point are put to use in understanding how a shell and tube exchanger works. The expression below is used consistently to quantify the physical conditions that effect the performance of a shell and tube heat exchanger.

$Q = U A \Delta T$ Q = Heat Load BTU/Hr

U = Overall heat transfer coefficient BTU/hr/Ft^2

ΔT = Mean average temperature difference between the process and the heat source or coolant.

Actually, a number of factors impact on the heat transferred through a wall such as thermal conductivity, surface films, and wall thickness. The overall coefficient, (U) is an effective tool to combine all the individual resistances into one overall transfer coefficient.

SHELL AND TUBE - THE WORKHORSE

A.   General Description of a Cooler/Heater

The shell and tube heat exchanger is a bundle of tubes connected together at each end in plates called tube sheets (See Figure 1). The tube sheet provides the seal between the tube bundle and the shell, which isolates the process from the service. It is very important that the seals are leak proof to prevent cross contamination of the fluids on the tube and shell sides. The basic shell and tube heat exchanger serves equally well as a cooler, heater, evaporator, or condenser with minor design changes in shape and arrangement.

The shell is a cylinder into which the tube bundle is housed. The ends of the shell are sealed with plates called heads. The head can be flat or dished.

Nozzles are installed in the heads for the inlet and exit flow through the tubes. Nozzles are also installed on the shell for the in and out flow on the shell side. There may be additional nozzles for drains, vents, and instruments.

The shell and tube heat exchanger meets several important process requirements.

1. The flow through the tubes is isolated from the flow through the shell.

2. The material used to construct the shell is compatible with the chemical in the shell and the tube material is compatible with the chemical through the tubes. Corrosion is the usual concern.

3. The tubes can be operated at a much higher pressure than the shell for the same thickness, (Hoop stress is a function of diameter).

4. Tubes are an efficient way to install a large amount of heat transfer area into relatively small amount of shell space.

5. Baffles can be installed along the tube bundle. The purpose is to lengthen the flow path in order to increase velocity, which improves mixing and convection. Additional tube support is a second benefit.

6. The heads are usually flanged to the shell. This permits easy access for inspection and maintenance of the tube bundle.

7. The heater/cooler exchanger usually performs well when installed either horizontally or vertically.

B. Reboiler

A modified shell and tube heat exchanger is often used as a reboiler, calandria, or vaporizer. The operating principle is to vaporize a liquid and use the latent heat in the process to do useful work. Establishing the vapor flow (boil-up) in a distillation column is a typical example.

The heat load Q on a reboiler is very large due to the latent heat (heat of vaporization). To keep the transfer area down so that larger tubes can be used (less pressure drop) high pressure steam is used as the heat source (high ΔT). It is also common practice to use a circulation pump on viscous fluids to improve heat transfer by convection through improved mixing (See Figure).

The reboiler may have a characteristic "hump" which provides vapor-liquid disengaging space.

C. Condenser

The shell and tube exchanger is also an effective condenser and is usually mounted vertically. The vapor, as it condenses, runs down the tubes into an enlarged base or perhaps a head tank, which maintains a seal on the system. Non-condensable are

vented to maintain system pressure. Steam jets are frequently used to control pressure in vacuum systems by removal of the non-condensable.

CONDITIONS THAT AFFECT HEAT TRANSFER

A. Mixing

Good dispersion of the heat being transferred (convection) to and from both streams is important for consistent performance and good control of the driving force (ΔT). Two ways good mixing is obtained by the use of baffles and high flow velocity.

Adequate flow velocity is determined by use of the Reynolds number.

Reynolds number $N_{RE} = \dfrac{DG}{M}$ D = Tube diameter
G = Mass velocity
M = Viscosity

The Reynolds number is used as a measure of the flow required to get good mixing as the density (Wt. Per unit volume) and viscosity (resistance to flow) vary. For this course, it is necessary to only understand its use.

B. Viscous Fluids

Heat transfer is more difficult into and out of viscous fluids. They mix poorly which reduces the heat transferable by convection. They adhere to the wall more strongly which increases the film resistance. One solution is to keep the Reynolds number as high as pressure drop permits which can be accomplished by external circulation.

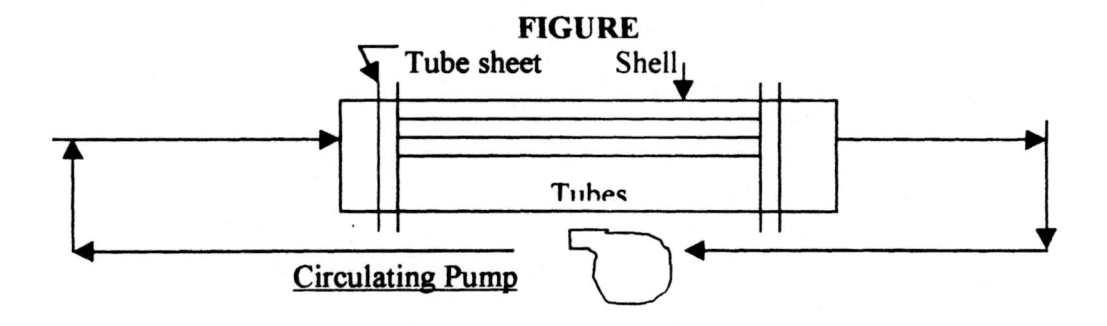

FIGURE

C. Transfer Area

Area is increased by using smaller diameter tubes in the same space. One two-inch diameter tube can be replaced by four one-inch tubes and the area double.

However, the velocity increases in small tubes which increases pressure drop. As a general rule, velocity in the tubes in or close to the design range is important for good performance.

D. Evaporation

When liquids evaporate on a wall, bubbles form. As the bubbles break free, the agitation (boiling) in the liquid can actually improve heat transfer. However, if the driving force ΔT is too high, the wall becomes coated with bubbles, which act like an insulation. Heat transfer efficiency is reduced.

E. Arrangement

Most shell and tube coolers and heaters are installed horizontally to simplify maintenance work such as tube cleaning or bundle replacement. However, in some cases it is important that the exchanger be mounted vertically where arrangement (elevation, close coupling and etc.) and vapor liquid separation are important.

E. Reboiler

In the reboilers, close coupling is important to reduce pressure drop since flow is created by density gradients (temperature difference) between the liquid in the process vessel and the liquid being heated in the boiler. The circulating line is large to reduce pressure drop and loaded to allow for thermal expansion.

The elevation of the vapor return is set to feed the vapor directly to the point it is needed in the process vessel.

Close coupling of a condenser is not as critical but the vapor piping is commonly quite large to reduce pressure drop. A more important consideration is liquid/vapor separation.

As the vapor condenses it drains down the tubes into a sealed tank. The upper section of the exchanger operates essentially dry until the vapor is cooled to the dew point. Non-condensable are vented at the bottom which maintains a flow in contact with all the tube area and can be used to control the operating pressure.

SPECIAL EXCHANGERS

A. Air Cooled

Where water is scarce and the process temperature is relatively high, an air-cooled exchanger may be practical. It is, in principle, a radiator not too different in principle from the type used to cool engine oil.

Simply, this unit is a tube bundle across which air is pushed or pulled by fans. Some of the fans may be reversible to improve control in cold weather. The tubes are often fitted with fins to increase the heat transfer area.

B. Electric Heaters

Electric heaters are used for small heat loads in inconvenient areas. Aluminum is commonly used as process material. The transfer of heat is poor compared to liquid – liquid exchange in the shell and tube. Electric heat can be a reasonable choice where use of steam is not economical and the heat load is low. The advantage is simple connections and excellent temperature control.

C. Tracing

Freeze protection on piping is often necessary to prevent freezing of polymers and other materials that are normally solids. This protection is often provided by tubing steam heated wrapped around the pipe and thoroughly insulated. Steam flow through the tubing replaces heat lost to the atmosphere.

D.   Jackets

On vessels where heat removal is desirable or heat loss is undesirable, a jacket is used. Tubular reactors are often jacketed to transfer into cooling water. Some compressors and pumps may have a jacket to remove heat of compression.

BASIC CONTROL LOGIC (See Figure)8

The purpose of most heat exchangers is to control process temperature. That control may facilitate evaporation, condensation, crystallization, solubility, reaction rate, and other process needs.

In the cooler situation, coolant flow is usually controlled by the process temperature exit the cooler. Coolant flow is metered to troubleshoot exchanger problems such as fouling or poor mixing.

On a heater, steam pressure is usually controlled by the process temperature exit the heater. Steam flow is metered to troubleshoot trap function and heat load.

On condensers, non-condensers are usually vented to control pressure. Coolant flow is adequate to condense all the vapor to liquid at that boiling point plus cool the liquid to a practical storage temperature.

Steam flow on a reboiler is often used to control vapor rate or boil-up. Pressure drop is often a convenient way to monitor vapor rate.

SAFETY CONSIDERATIONS

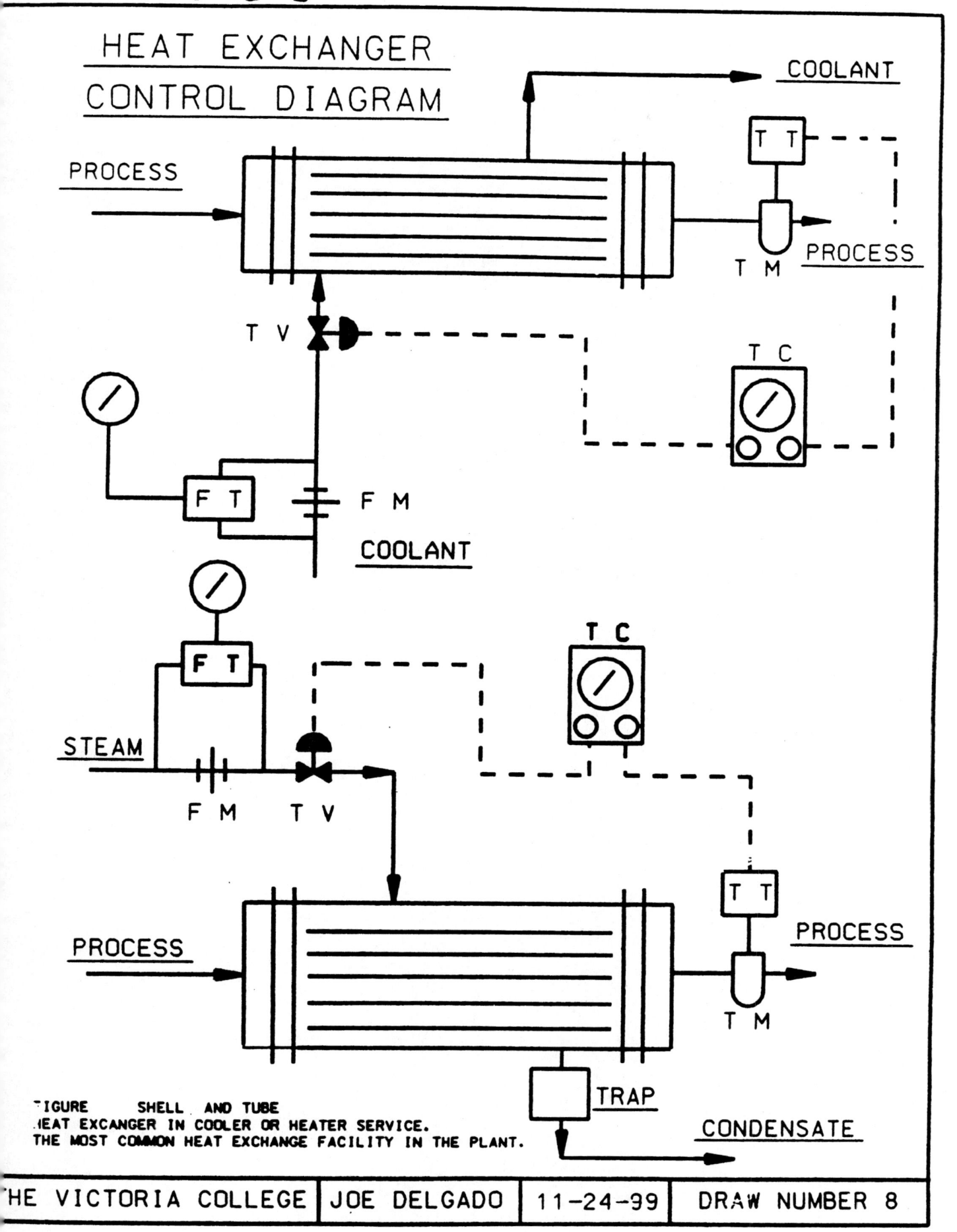

FIGURE 3

HEAT EXCHANGER CONTROL DIAGRAM

FIGURE SHELL AND TUBE
HEAT EXCANGER IN COOLER OR HEATER SERVICE.
THE MOST COMMON HEAT EXCHANGE FACILITY IN THE PLANT.

| THE VICTORIA COLLEGE | JOE DELGADO | 11-24-99 | DRAW NUMBER 8 |

A high-pressure steam leak can be invisible and the noise it makes should be approached cautiously.

Solids and liquids expand and contract at a predictable rate when heated or cooled. The expansion or contraction is equal in all directions in a system without restraint. The volume that cannot increase with a temperature change can generate a damaging pressure.

In a liquid filled system with no vent, the expansion is completely restrained and the pressure can build up or fall off rapidly with small changes in temperature, because liquids are essentially non-compressible.

The pressure generated by thermal expansion can rupture pipe, blow seals or gaskets, and crack vessel walls. In liquid filled systems, some type of pressure relief is usually provided. A relief valve or rupture disc is commonly used. These devices are for the protection of personnel and equipment from excessive pressure and should not be used as routine operating tools.

A condensing gas in a closed system creates a vacuum. The vacuum, in turn, creates an external force that the equipment must be designed for or there can be collapse at the weakest point. For example, each pound per square inch of vacuum creates 144 pounds of external force against each square foot of external surface wall.

It is not unusual for the materials on the shell side and tube side of a heat exchanger to be different to obtain acceptable corrosion rates. It is important that approved materials be used to replace any part of the exchangers. Seals are especially vulnerable to corrosion.

CHAPTER 3

COMBUSTION AND FURNACES

INTRODUCTION

Combustion is a term used to describe the burning of carbon based fuels under controlled conditions with the intent to use the heat that is released to do useful work. A better term to describe burning with the intent to destroy the material would be incineration.

Combustion is used in petrochemical plants to release large amounts of heat energy at high temperatures. The combustion is carried out in an insulated furnace (firebox) to minimize heat loss and provide sufficient time for transfer of as much of the heat of combustion as possible to a medium where the heat is used in industrial processes. Heat absorbed by water to produce steam followed by use of the steam in heat exchangers would be a good example.

Incinerators and flares utilize flame temperature to destroy (waste) materials, which either burn or decompose at the flame temperature. Often the fuel itself is the material to be destroyed as a safe means to dispose of a flammable gas or liquid waste that would contribute to air pollution or pose a fire hazard if released untreated.

The high temperature of combustion can also be used to crack or break down large molecules into smaller, more usable products. Crude oil cracking is an example. Ethylene plants use the same principle to break down a saturated hydrocarbon feedstock like propane or distillate into small molecules.

In order to conserve the huge amount of energy (and investment) involved, the combustion process has been extensively studied for many years. The two areas of most

interest are maximizing the heat release from the fuel and optimizing the transfer of the heat of combustion into a medium to do useful work. Overall conservation of heat in steam generation is a major industrial objective. The chemical reactions in the combustion process are among the most basic found in industry. However, the technology at work is common to all chemical reactions.

This study is limited to the use of heat energy released by combustion to generate steam – one of the more complicated of heat transfer systems with radiation, convection, and conduction all playing a major role in overall efficiency.

CHEMICAL AND PHYSICAL PRINCIPLES

This lesson states with the basic chemical reactions and proceeds to actual furnace conditions. Energy and material balances are used to explain the chemical and physical changes that take place.

There are two basic chemical reactions involved in all industrial combustion processes.

REACTANTS		PRODUCTS	
Hydrogen + Oxygen	→	water	
$2H_2 + O_2$	→	$2H_2O$	Atomic balance
4 lbs. + 32 lbs.	→	36 lbs.	Weight balance

The equation reads: Two moles of hydrogen combine or react with one mole of oxygen to form two moles of water. Using the atomic weights, H = 1 and O = 16, it follows that as 4 lbs. of hydrogen react with 32 lbs. of oxygen to produce 36 lbs. of water. Notice that in a balanced equation the atoms and weight in equals the atoms and weight out.

The second basic chemical reaction is:

REACTANTS		PRODUCTS	
Carbon + Oxygen	→	Carbon Dioxide	
$C + O_2$	→	CO_2	Atomic balance
12 + 32	→	44	Weight balance
44 lbs.			

The equation reads: One mole of carbon reacts with one mole of oxygen to form one mole of carbon dioxide. Using the atomic weights, C = 12 and O = 16, it follows that 12 lbs. of carbon react with 32 lbs. of oxygen to produce 44 lbs. of carbon dioxide. Again, the number of atoms in the reactants equals the number of atoms in the products and produces a weight balance.

Pure hydrogen and carbon are rarely, if ever, available as industrial fuels. However, chemical compounds of carbon and hydrogen such as natural gas (NG) and coal are plentiful. Since natural gas is essentially composed of methane (CH4) and is so widely used as an industrial fuel, it is used as the reactant in the study of combustion.

Chemical reactions are studied in the laboratory using grams and calories as the basic units of weight and heat with degrees centigrade °C as the standard unit of temperature. In most chemical plants, the weight is in pounds, heat energy is in BTUs, and temperature is in degree Fahrenheit °F.

The Balanced Equation for the <u>Combustion of Methane (NG)</u>

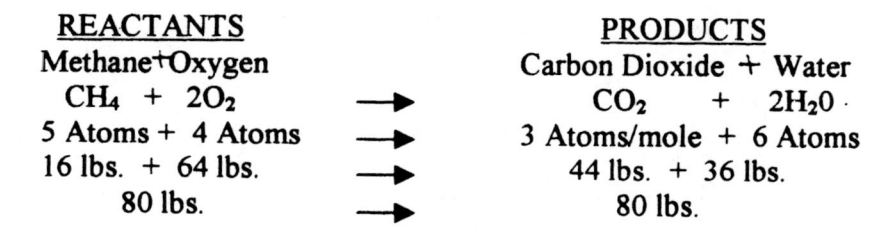

REACTANTS		PRODUCTS
Methane+Oxygen		Carbon Dioxide + Water
$CH_4 + 2O_2$	$\rightarrow$	$CO_2 + 2H_2O$
5 Atoms + 4 Atoms	$\rightarrow$	3 Atoms/mole + 6 Atoms
16 lbs. + 64 lbs.	$\rightarrow$	44 lbs. + 36 lbs.
80 lbs.	$\rightarrow$	80 lbs.

The molecular quantities and weights that balance the equation are the exact stoichiometric quantities that react to produce products. The equation is based on 100% conversion of reactants to products and 100% yield to products that yield the most energy. Such an ideal reaction is unlikely, if not impossible, to attain in industrial processes. However, the equation is useful to demonstrate energy and material balances in their most simple form.

ENERGY BALANCE

The basic principle upon which the heat of reaction (combustion) is calculated is stated as:

The difference between the <u>sum of</u> the heats of formation of the products and the <u>sum of</u> the heats of formation of the reactants is equal to the heat of reaction (combustion).

By thermodynamic convention, the heat of formation, which is the energy needed to form a compound from its elements, is given a negative value. Elements are given a zero heat of formation. Using this convention, exothermic reactions have a – (negative) heat of reaction and endothermic reactions have a + (positive) heat of reaction.

The standard heat of combustion of methane (CH_4) is defined as the change in enthalpy of reactants and products at 77°F from the oxidation reactions. To be more specific, consider the oxidation of methane in air under stoichiometric conditions. The reaction is as follows:

$$CH_4 + 2O_2 + 7.52 N_2 \longrightarrow CO_2 + 2H_2O + 7.52 N_2$$

Usually stated in cal./gm - mol

Using heats of formation, we can calculate the enthalpy of products and reactants at 77°F. The heat of combustion for this reaction is the change of enthalpy of the reactants and products as follows:

                        CO_2 H_2O N_2

ΔH Products = 94,052 cal/gm-mol + (2) (-57,798) + 0

               = -209,648 cal/gm mol

to convert to BTU/lbs. mol, multiply by 1.8

               = -377,366 BTU/lb.-mol

                        CH_4 O_2 N_2

ΔH Reactants = -17,889 cal/gm-mol + (2) (0) + (7.52) (0)

ΔH Reactants = -17,889 cal/gm-mol x 1.8

               = -32,200 BTU/lb.-mol

ΔH Combustion = -377,366 - (-32,200) BTU/lb.mol

$$= -345{,}166 \text{ BTU/lb.-mol}$$

$$= 21{,}519 \text{ BTU/lb. of CH4 combusted}$$

$$= \sim 1{,}000 \text{ BTU/FT}^3 \text{ combusted}$$

TABLE 1

PROPERTIES OF MATERIALS INVOLVED IN

THE COMBUSTION OF METHANE

Oxygen, O_2

Mol. wt.	32
Specific heat, (Cp), BTU/lb./°F	0.22
Heat of formation, BTU/lb. mol.	0.0

Methane, CH_4

Mol. wt.	lb.
Specific heat, (Cp), BTU/lb./°F	0.53
Heat of formation, BTU/lb. mol.	-32,200

Carbon Dioxide, CO2

Mol. wt.	44
Specific heat, (Cp), BTU/lb./°F	0.30
Heat of formation, BTU/lb. mol.	-169,380 BTU/lb.mol.

Water, H_2O

Mol. wt.	18
Specific heat, (Cp), BTU/lb./°F	0.55
Heat of formation, BTU/lb. mol.	-104,040

Nitrogen, N2

Mol. wt.	2.8
Specific heat, (Cp), BTU/lb./°F	0.27
Heat of formation, BTU/lb. mol.	0

Composition of Air	Nitrogen	Oxygen	Other
Vol. Or mol. %	78.1	21.0	0.9
Wt. %	75.5	23.2	0.3
Air specific heat (Cp), BTU/lb./°F			0.26

Materials that are present but do not take part in the chemical reaction are called inerts. Nitrogen, then, is an inert in the combustion process and goes through unchanged. However it does absorb heat energy. This lowers the flame temperature and ultimately increases heat lost in the stack gas.

The balanced equation with nitrogen:

Reactants	Inert		Products	Inert
$CH_4 + 2 O_2$	$+ N_2$	$\rightleftharpoons$	$CO_2 + 2H_2O$	$+ N_2$
16 + 64	+ 212	$\rightarrow$	44 + 36	+ 212
292		Total Pounds	292	

64 ÷ 0.232 = 276 lbs. of air in
276 - 64 = 212 lbs. N_2 in and out

0.232 is the weight fraction of oxygen in air. All other components are lumped together as nitrogen.

The moles and pounds in the equation are stoichiometric quantities or the exact amounts to produce the products form the reactants. This perfect reaction is not practical. In practice, the law of mass action is used to adjust reactant concentrations in order to make a reaction go in the direction that gives the best results. A common desired result in combustion is to efficiently consume as much fuel as possible with the maximum release of useful heat. This objective is accomplished, typically, by use of excess oxygen which is provided from excess air (10% excess is typical but not mandatory).

The combustion equation with 10% excess air:

Reactants	Inert	Excess		Products	Inert	Excess
$CH_2 + 2O_2$	$+ N_2 +$	O_2	$\rightleftharpoons$	$CO_2 + 2H_2O$	$+ N_2 +$	O_2
16 + 64	+ 233.2	+ 6.4	$\rightarrow$	44 + 36	+ 233.2	+ 6.4
	319.6		Total Pounds		319.6	

<u>Oxygen and Nitrogen Balance</u>

Excess O2   64 + (0.10)(64) = 70.4 - 64.0 = 6.4
Excess N2   212 + (0.10)(212) = 233.2 - 212.0 = 21.2

In conclusion, the efficient combustion of 16 pounds of methane produces close to 320 pounds of products at around 2000°F

The Actual Energy Balance

The information is now in place to complete an actual energy balance. The heat release as calculated from the heat of formation is 345,000 BTU for each 16 pounds of methane burned. With the use of 10% excess air to efficiently combust the gas, (oils do better at 15% excess air), 320 lbs. of reactants and inerts produce 320 lbs. of products and inerts. The important principle to remember is that all the heat released must initially be absorbed by the products, inert, and excess oxygen.

Reactants	Inert	Excess		Products	Inert	Excess
$CH_2 + 2 O_2 +$	$N_2 +$	O_2	→	$CO_2 + 2H_2O +$	$N_2 +$	O_2
16 + 64 +	233.2 +	6.4	→	44 + 36 +	233.2 +	6.4
	320 lbs. Total Pounds				320 lbs.	

Data are rounded to simplify the arithmetic. The specific heat, (Cp) of the products is taken as 0.5 BTU/lb./°F. The reference temperature is 77°F. The flame temperature is assumed to be 2100°F.

Step 1

The 320 lbs. of reactants produced 320 lbs. of products with an increase in temperature of 2100°F - 77°F or 2023°F.

The increase in heat content ΔH (enthalpy) of the products including the inert nitrogen and excess oxygen has increased and can be calculated.

$$\Delta H = (mass)(Cp)(\Delta T)$$ ΔT Temperature difference, °F

$$\Delta H = (320)(0.5)(2023)$$ Cp specific heat, BTU/lb./°F

$$\Delta H = 324,000 \text{ BTU}$$

For more rigorous calculation, we would use the sum of the heat content of each component i.e.

(CO2, lbs.)(Cp of CO2)(ΔT) + (H2O, lbs.)(Cp of water)(ΔH) and etc. = ΔH

However, the 324,000 BTU compares favorably with the theoretical maximum of 345,000 BTU heat of reaction.

Step 2

The 324,000 BTUs in the products goes to two places in a steam generation plant.

1. Sensible heat to heat the feed water and latent heat to generate steam.
2. Discharge of the stack (flue gas) which takes heat energy out of the system.

A typical stack temperature is 800°F. The heat transferred to the process is:

ΔH = (mass)(Cp)(ΔT)   2100°F is the assumed flame temperature

= (320)(0.5)(2100 – 800)

= 208,000 BTU

The thermal efficiency = $\frac{208000}{324000}$ x 100 = 64%

As a check, the heat content of the stack gas can be calculated:

ΔH = (320)(0.5)(800 – 77)

= 115,500

208,000 + 115,500 = 323,500 which compares favorably to the 324,000 BTU/lb. calculated above.

With the sensible heat plus the heat of vaporization of approximately 1000 BTU/lb., the heat from 6 pounds of methane combusted will produce:

208,000 BUT ÷ 1000 BTU/lb. = 208 lbs. of steam for each 16 lbs. of methane.

THE BASIC FURNACE

A furnace is an enclosed space where a fuel is burned under controlled conditions referred to as combustion. The products of the combustion called the stack or five gas are discharged through a stack to the atmosphere. If the furnace is used as a vent source to do further work such as steam generation, the combustion chamber is often called a firebox.

Fuel into the burners is burned completely before passing out the flue duct to the stack. The firebox is lined with special bricks or a refractory so it can withstand high temperatures without melting or spalling. The firebrick or refractory and additional insulation prevents heat loss to the outside atmosphere and protects the structural steel in the furnace from excessive temperatures.

A typical gas burner feed line is equipped with an eductor or aspirator, which sucks air into the fuel gas fed to the furnace. The amount of air sucked in is controlled by the opening of an adjustable shutter. This shutter is called the primary air register. The air that is sucked in is mixed with the fuel gas in the mixing tube on the way to the burner portholes. for this reason, air and gas are "pre-mixed" and the burner is called a "pre-mix" burner. The mixing of air and hydrocarbons before reaching the burning zone improves combustion and has the advantage of a shorter flame, which is a desirable feature of a burner. All of the air needed to support combustion cannot be supplied in the pre-mixing section. The rest of the air needed is sucked by draft through a secondary adjustable shutter. This is called secondary air. A desired burner is one that can handle a large volume of fuel gas while providing a short clear flame so that high heat liberation will be contained in the firebox. Such a situation can only be achieved by thoroughly

mixing air with the gas prior to combustion and providing additional or secondary air to the burning zone to endure complete combustion in the immediate burner area.

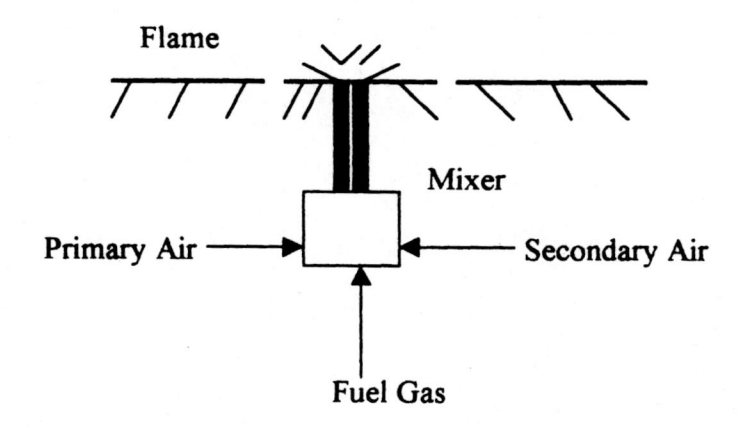

FURNACE DRAFT

The draft at any location within a furnace is the difference between the pressure at that location and the pressure of the atmospheric air on the outside. This draft, or difference of pressure is caused by the difference between the weight of the vertical column of hot flue gas in the furnace stack and the weight of a column of cooler outside air of the same height. The weight is directly proportional to density, which is proportional. The cooler outside air is heavier. As it contacts the air openings around the furnace burners, its greater weight causes it to rush through these openings and push the lighter, hotter flue gases up the stack. The cool air is heated instantly in the furnace, then it is pushed up the stack by more cool, heavier air. In this manner the movement of air through the furnace becomes continuous.

If the volume of flue gas to pass through the system is large, the stack and duct must be made large enough to accommodate it with very little pressure drop. Since the stack is usually designed for maximum conditions, there may be too much draft for proper firing under normal conditions and a draft control system must be provided by mechanically obstructing the movement of flue gas up the stack. Control of this natural draft is accomplished by installing a damper in the breaching or the stack.

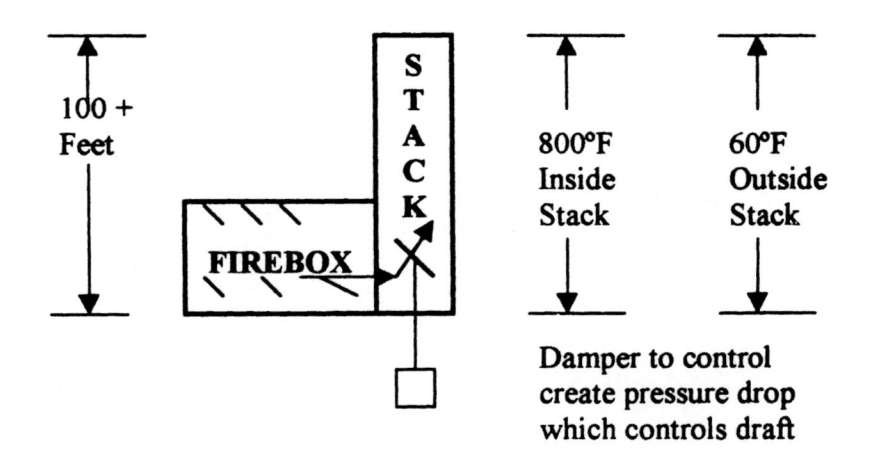

FLAMMABILITY AND EXPLOSIVE LIMITS

Combustion only occurs within a concentration range of a fuel air mixture called the flammability limits. The oxidation of combustibles is still viable outside the flammability range using a catalyst to modify reaction conditions. Lowering of oxidation temperature is a common way to produce alcohol and ketones. The flammability limits for methane (CH_4) are 5.0 to 15.0 volume % methane in air.

A mixture of fuel in air also has explosive limits. Most chemical plants go to a considerable effort to insure that explosive limits of process materials are never reached. The explosive limits of methane in air are 5.3 to 13.9 volume %.

One of the most flammable and explosive of all commonly used plant chemicals is hydrogen. With an explosive range of 4.1 to 74.2 vol. %, absolutely no risk of ever mixing hydrogen and air should be allowed. Other common chemicals with a wide explosive range that deserve special precautions are hydrogen sulfide (4.3 – 46 vol. %), acetylene (2.5 – 80 vol. %), methyl alcohol (6 – 36.5 vol. %), and ethylene (2.8 – 29 vol. %).

A common cause of a furnace explosion is ignition of a rich mixture in the firebox where there is too little air for normal. The most common cause of a rich mixture and explosion is loss of ignition. For example, if the fire becomes unstable and goes out, the fuel entering the furnace will build up around the tube surfaces. If ignition is then re-established an explosion will occur. Whenever ignition is lost on a boiler, the fuel supply must be cut off immediately. The rich mixture is purged from the boiler by a large amount of air. Flame scanners on each burner protect against flame out. The flame scanner detects the loss of flame and immediately shuts off the fuel supply. It will also

prohibit individual burners from being fired until the igniter and ignition burner is working properly.

During a short shutdown, combustible gases may build up inside the furnace if there is leakage through the shutoff valves. Before trying to light a burner, the furnace must be purged completely with a large amount of air for at least five minutes. This is known as the purge cycle. The purge relay will not permit any fuel supply to the boiler until the purge cycle has been completed.

TUBE RUPTURE

Another serious safety concern is the sudden injection of liquid water into the firebox from a ruptured tube or a tube connection. Each pound of water flashes to produce 20 cubic feed of steam almost instantaneously. Though not a true explosion, the flash of water into steam with the accompaniment by a pressure surge produces similar results.

Pressure boilers are among the most frequently and rigorously inspected of all process equipment. Corrosion, scale buildup, and performance of safety devices are typical checks thoroughly inspected and recorded.

TERMS AND SYMBOLS

<u>Convection Section</u> - The part of a furnace between the radiant section and the stack. This area is filled with tubes or pipes which carry a process stream and which absorb heat via convection heat transfer from the hot gases passing through the area on their way out the stack. The convection section forms an obstacle to the combustion gas flow and can greatly affect furnace draft in the radiant section of the furnace.

<u>Radiant Section</u> - The part of a furnace into which the burners fire. Tubes mounted in this area of the furnace receive heat principally via direct radiation from both burner flames and furnace refractory. Physical volume and arrangement of the radiant section has a great affect on burner choice and required flame patterns.

<u>Burner</u> - A device which combines fuel and air in proper proportions for combustion and which enables the fuel-air mixture to burn stably to give a specified flame size and shape.

<u>Combustion</u> - The rapid oxidation of a substance accompanied by the release of energy and light. It can be described as fire under controlled conditions.

<u>Heat of Combustion</u> - The energy released by the chemical reactions in a fire. It is the sum of the heats of formation of the products minus the heats of formation of the reactants.

<u>Flame Temperature</u> - The temperature reached by the oxidation reaction or combustion within the flame or light-producing zone. The flame temperature varies with the type and quality of the fuel but hovers around 2000°F.

<u>Ignition Temperature</u> - Not of much practical use except as a guide to how easily a substance is to ignite. Also, any release of a substance at a temperature above its ignition

temperature will most likely ignite and start a fire. A number of industrial processes operate above the ignition temperature.

Flame Propagation - The consumption of gaseous fuel (burning rate) against the flow of fuel from a burner. In furnaces, the air is usually premixed with air. If the flow of fuel to a burner higher than the flame propagation rate the flame will "blow out".

Flash Point - The temperature of a flammable liquid where the vapor above the liquid reaches the ignition temperature.

Flammability Limits - The upper and lower concentrations of a fuel in air that will sustain combustion. A high concentration of fuel is a rich mixture and a low concentration of fuel is a lean mixture. Neither too rich nor too lean combusts (burns) well.

Explosive Limits or Explosive Range - The leanest and richest mixture of fuel in air than can explode if ignited.

Enthalpy - This is the term given for the total energy, due to both the pressure and temperature, of a fluid or vapor (such as water or steam) at any given time and condition.

The basic unit of measurement for all types of energy is the British Thermal Unit (BTU).

Specific Heat (Heat Capacity) - A measure of the ability of a substance to absorb heat. It is the amount of energy (BTU's) required to raise 1 lb. by 1°F. Thus specific heat capacity is expressed in BTU/lb.°F. The specific heat capacity of water is 1 BTU/lb.°F. This simply means that in increase in enthalpy of 1 BTU will raise the temperature of 1 lb. of water by 1°F.

Excess Air - A convenient term used to define the amount of oxygen present over the stoichemetric amount needed to completely combust (oxidize) the fuel.

Stoichiometry - The area of chemical reactions that defines the exact amount in which reactants produce products. All chemical equations are written and balanced in stoichiometric amounts. The number of atoms in the products is equal.

Reactant - The original chemicals that take part in a chemical reaction.

Products - The chemicals produced from the reactants.

Inerts - Chemicals that are present during the reaction but take no part.

Heat of Formation - The heat absorbed or evolved when elements join to form a compound. This heat energy valve is tabulated for most known compounds at a reference temperature. A common reference is 25°C (77°F).

Heat of Reaction - The heat energy absorbed (endothermic) or evolved (exothermic) during a chemical reaction. Thermodynamically, the heat quantity is the difference between the sums of the heats of formation of the products and reactants.

Molecular Weight - The sum of the atomic weights in a chemical compound

Enthalpy - The heat content of a substance, which includes both latent and sensible heat, from absolute °0 to T°. Changes in enthalpy (ΔH) of a substance accompanied by a temperature change (ΔT) is more useful in calculating the heat in processes.

Draft - The slight negative pressure in the firebox due to the density difference between the hot gases in the stack and an outside column of air of equal height.

Stack Gas - The combustion products vented to the atmosphere. Also, called flue gas.

Furnace - An enclosed area designed to carryout combustion. It is also called a firebox and a combustion chamber.

CHAPTER 4

STEAM GENERATION

INTRODUCTION

The production of a totally dependable supply of high quality steam is a key industrial process in the operation of a petrochemical plant. Steam generation is also a significant process because of the large amount of heat energy generated and consumed continuously. Condensate recycle is now standard practices in the operation of ever more efficient steam generation facilities.

Design and construction of steam boilers has been a major business for engineering firms for decades. Few operations have received as much attention with two prime objectives:

1. Efficient use of energy (thermal efficiency)
2. High on line availability of high quality steam.

The steam boiler is the first unit on line and the last to be shutdown. A complete shutdown of steam generation and supply is a rare event and has serious effects if not carefully planned.

High boiler utility is primarily obtained by careful control of combustion conditions and a rigid maintenance program. The use of wastes as fuel in commercial boilers is a fairly recent innovation and has eliminated wasted energy in flares and incinerators. Wastes are destroyed in a flame and the heat is used to produce steam. If the waste stream is of reasonable quality (clean, good fuel value, dependable source and etc.) it is used in the main boilers through use of special or modified burners. Wastes are usually a small percentage of the total fuel demand and the dilution is good protection

against major upsets. If a waste stream is of poor quality as a fuel, a special boiler may be dedicated to its use. Low-pressure steam is normally produced in this special equipment which is referred to as a waste heat boiler. The waste stream is not always a fuel. It may be an oxidizing agent and support combustion of the fuel. The high temperature in the firebox may be used to de3story an undesirable chemical before it is vented directly into the atmosphere.

Steam generation occurs in a complex of operations called a powerhouse. The powerhouse is a logical location as the custodian of other highly reliable systems, which may include:

1. Plant air – instrument air
2. Backup fire water
3. Incoming utilities such as electrical power and natural gas
4. Water treatment
5. Inert gas

STEAM

BOILER CONFIGURATION

The steam and water drum or upper drum is the vessel at the top of the boiler (See Figure 1). This drum is where the feed water to the boiler enters and where the level of water is observed and maintained. Also, it is the vessel that receives the steam from the generating tubes or risers and directs this steam into the main header. There is an internal feed line that distributes the feed water. There is also an internal chemical feed line for use where chemical treatment is necessary.

The mud drum or lower drum receives its water from the upper drum via the down comers. The water going up through the risers pick up heat and is turned into steam. The lower drum has a connection at each end of the drum. These are used to purge sediment called blowdown during continuous operation and for draining the boiler when it is out of service.

The down comers are those tubes that are located in the cooler section of the boiler used to transfer the water from the upper drum to the lower drum. The risers are by far the largest tube section in the boiler. They take water from the lower drum where heat is added to the steam and water in the upper drum.

FIGURE 1. TYPICAL BOILER COMPONENTS

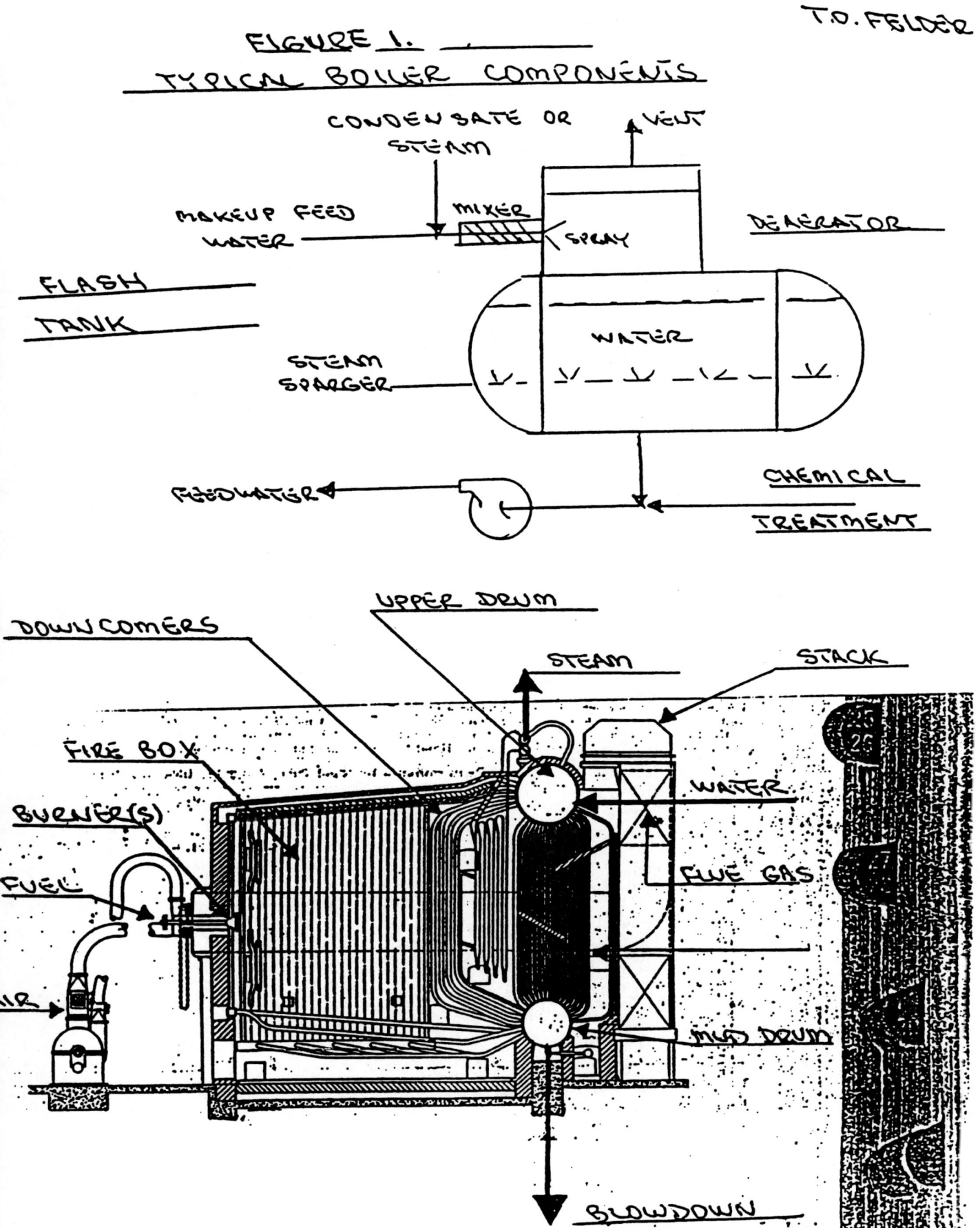

OPERATING PRINCIPLES

As heat is added to water, the vapor pressure increases. When the vapor pressure reaches the surrounding pressure, the water is said to be saturated. Any additional heat causes boiling and vaporization if the pressure remains constant.

Consider a boiler as a closed vessel. As more steam is generated it is compressed and exerts a pressure on everything surrounding it. This includes exerting a pressure on the surface of the water. If the pressure on the surface of the water increases, the temperature of saturated water increases. While at atmospheric pressure the temperature of saturated water is 212°F. At a pressure of 150 psia, the temperature of saturated water is 358°F (See Figure). So the temperature and pressure of steam have a fixed relationship. One sets the other and the quantity of steam produced is controlled by the heat input.

Figure

Heat Content of Steam

212°F	Latent Heat 970.6 BTU/lb	Total Heat 1150.8 BTU/lb
212°F	Sensible Heat 180.2 BTU/lb	
32°F		

Total Heat content of steam at atmospheric pressure

358°F	Latent Heat 864.5 BTU/lb	Total Heat 1195.1 BTU/lb
358°F	Sensible Heat 330.6 BTU/lb	
32°F		

Total Heat content of steam at 150 psi abs pressure

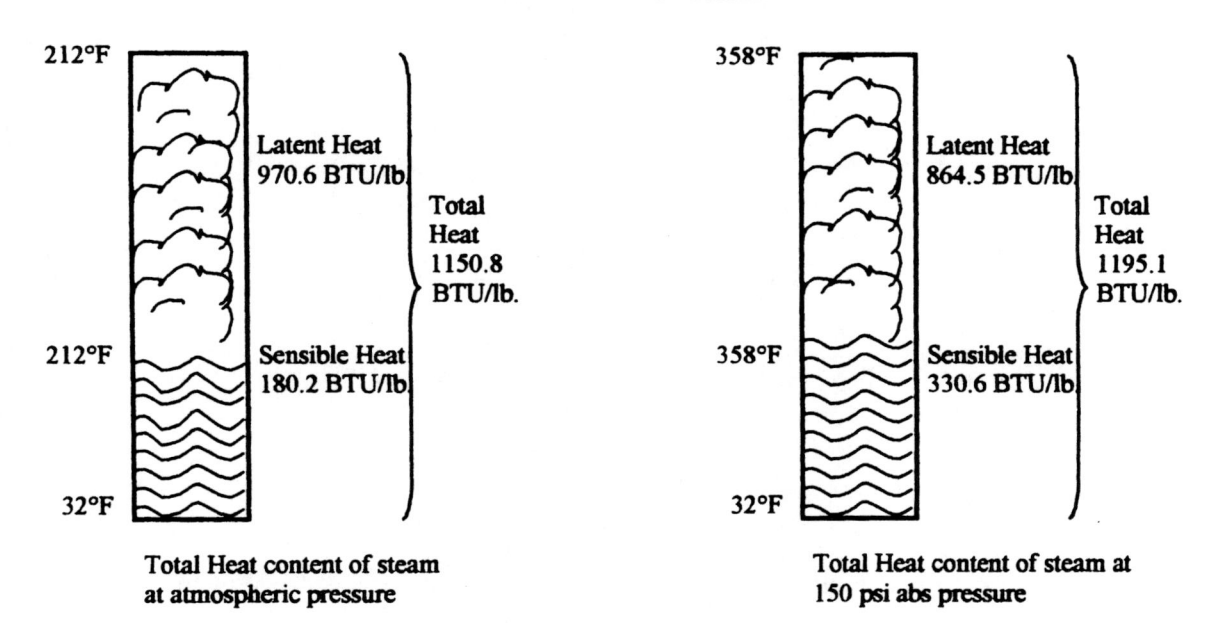

Another interesting observation is that the total heat energy in steam is fairly constant but a significant shift from latent to sensible heat occurs as pressure and temperature increases.

Steam is produced from boiler feed water that is a combination of treated water and steam condesate that is collected and returned from the individual industrial processes. Dissolved gases in the feed water may be corrosive as well as non-condensable in the steam distribution system. The problem is solved by aeration (See Figure 3). This process reduces gas solubility by increasing water temperature and stripping the gases from the heated water. Low-pressure steam injection is used for this purpose.

The boiler feed water is raised to the operating pressure of the boiler by a multi-stage pump. Adequate pump suction head is of major importance to prevent cavitation since the boiler feed water may well be close to the boiling point at storage conditions.

In modern boilers, the feed water is fed through an economizer (pre-heater). The heat transferred to the water in the economizer reduces the heat load on the vapor generating components of the boiler. This provides an increase in thermal efficiency.

The amount of sensible heat to raise the boiler feed water to the boiling point is significant. If the feed water is at 212°F going into a 600 psig boiler (489°F) the sensible heat per pound of steam is:

$$\Delta H = (M)(Cp)(\Delta T)$$
$$= (1)(1.0)(489 - 212)$$
$$= 277 \text{ BTU/lb.}$$

DEAERATOR

IS USED TO STEAM STRIP CORROSIVE GASES FROM BOILER FEED WATER

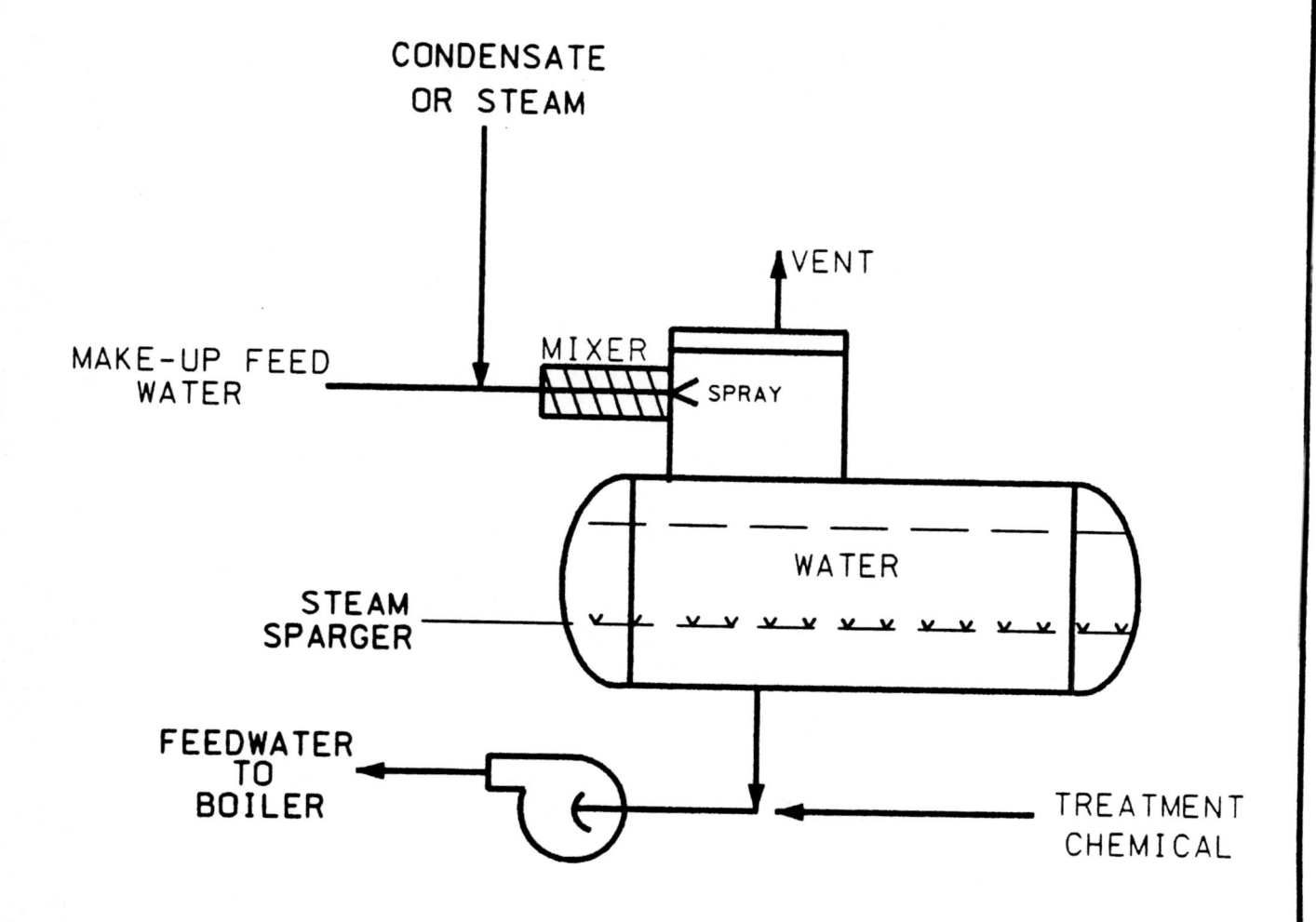

FIGURE 3

THE VICTORIA COLLEGE	FLOW DIAGRAM		
ADRIANA BURLEIGH	NTS	11-20-99	DRAW 8-B

The boiler feed water is fed into the top drum of the boiler. The saturated steam is produced by flashing (evaporation) in the top drum. This occurs at the boiling point of water at the operating pressure of the drum. Typical is 489°F to produce 600 psig steam.

Scale formation is a common problem so a purge of water from the mud drum is taken to control solids buildup.

Water circulates naturally through the down comers and backup the risers due to density gradients created by the temperature difference between the upper and lower drums – a thermosyphon effect. The steam generation facilities are exposed to <u>radiant</u> heat from the burners at temperatures up to 2100°F. Heat is being transferred through the tube walls at temperatures between 2100°F down to 800°F by <u>conduction</u>. The combination of gravity and density change keeps the circulation rate high, which produces good mixing and heat transfer by convection. The result is that all three heat transfer mechanisms must be efficiently controlled for economical steam generation.

The latent heat (heat of vaporization) of water at 489°F is 728 BTU/lb. The heat transferred in a 600 psig boiler to produce steam from 212°F feed water is:

$\Delta H_T = H_V + H_S$ H_V = heat of vaporization, BTU/lb.

$= 728 + 277$ H_S = sensible heat, BTU/lb.

$= 1005$ BTU/lb.

The steam from the top drum goes through a demister to remove any entrained water droplets. The steam is superheated slightly as it is fed into the distribution system. This superheat is controlled, if necessary by water injection and evaporation.

Heat increases the rate of corrosion. A major corrosion factor is acidity so pH is carefully monitored. Oxygen, the major factor in the embrittlement mechanism of corrosion is also carefully controlled.

TERMS AND SYMBOLS

<u>Stack Gas (flue gas)</u> - The products of the combustion process.

<u>Economizer</u> - A pre-heater for the boiler feed water.

<u>Superheater and Superheat</u> - Superheat is the amount of heat energy put into a vapor above the boiling point. The heat exchanger that provides the heat energy is called a superheater.

<u>Saturated Steam</u> - Pure steam at the temperature that corresponds to the boiling point temperature of the water at the operating pressure.

<u>Cavitation</u> - When the suction head on a pump is below the performance curve recommendation, the low pressure at the suction causes vapor formation by flashing. The water vapor, being compressible, contracts, and expands with fluctuations in pressure creating a noise called cavitation. In extreme, the pump "gases off" and heats up even more from the motor energy without the liquid flow to continually remove heat energy in the discharge. Even in mild cases, flow and pressure are erratic. The solution is a sufficient suction head (pressure) at all times.

<u>Risers</u> – The tubing between the top drum and mud drum where water flow (low density) occurs from mud drum to the top drum. Steam begins to form here from superheated water as the pressure falls.

Down comers – The tubing between the top drum and mud drum where <u>saturated water</u> (high density) flows down from the top drum into the mud drum due to the density difference.

Top Drum – The vessel where the superheated water flashes to steam.

Mud Drum – The vessel where water is superheated water flows through the risers to the top drum. Suspended solids called mud to settle out – thus the name mud drum.

Lowdown – A periodic or continuous removal of solids from the mud drum by discharge of water.

SAFETY

The boiler firebox (combustion chamber is essentially air tight and heavily insulated by a refractory to <u>enhance radiation</u> and conserve energy. However, the boiler walls are not designed for containment of pressure. In fact, the firebox operates at a slight vacuum (draft). The possibility of a fuel-air explosion was discussed under combustion and furnaces. Another potential explosion possibility exists if a major failure occurs in the water piping and spills a significant amount of water into the firebox.

Each pound of water produces 20 cubic feed of steam at atmospheric pressure and ambient temperature. At boiler firebox temperatures, that volume per pound of water increases three to four fold and occurs instantaneously. The firebox pressure increases rapidly until fire box failure occurs at the weakest point due to over pressure.

The pressure surge can happen so rapidly from the flash of water into steam that for practical purposes duplicates an explosion.

One major concern in any rigid system is a localized temperature change that creates a temperature differential, which results in a large thermal stress. As systems heat

up components expand to relieve stress. The possibility of damage is eliminated when that expansion occurs at the same rate (temperature rise) throughout the system.

In a boiler fire box, non-uniform heating creates temperature differences, which can crack refractory materials, damage tubing connections and etc.

The solution is to heat the fire box slowly on startup at a predetermined rate toward the operating temperature. The process may take several hours for a small boiler and up to 8 hours or more for a large one.

CHAPTER 5

STEAM DISTRIBUTION AND CONDENSATE RECOVERY

INTRODUCTION

Steam is distributed from the Powerhose to every area of a petrochemical plant. Dependable steam distribution is essential to most plant operations. That means, in practice, 100% utility is demanded from the steam supply and distribution system. The result is an unceasing effort to prevent equipment failure in all steam generation and distribution facilities. Most critical equipment has an installed spare as backup.

The large steam users are heat exchangers like process heaters, vaporizers, and reboilers. But even small users such as stem tracing, tank coil, and equipment jackets can be important to prevent plugged lines, freeze up of process streams, and to reduce heat losses. Steam is normally used at several pressure levels to meet temperature demands for different processes. This is accomplished by reducing stations installed along the main header that feed into area headers.

The main steam headers are usually mounted on pipe bridges designed to accommodate expansion and to handle their size and weight without sagging which is an important consideration for proper drainage. Condensation occurs due to heat loss to the atmosphere in all steam distribution systems because insulation is not perfect. The condensate has to be continuously removed since it reduces flow area, which increases pressure drop and increases header weight. Water hammer is also possible if a major flow change occurs which can accelerate a wave of water down the header.

Drip legs (collection pipes) are installed along the steam header to collect the condensate which is either drained or trapped for removal.

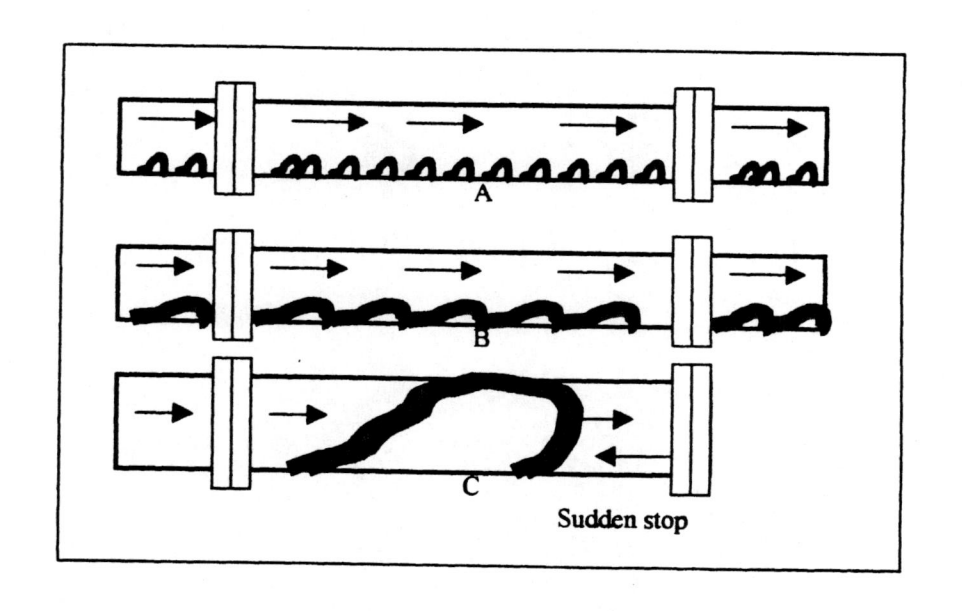

A

B

C
Sudden stop

PRINCIPLES OF OPERATION

A steam distribution system is essentially a series of closed loops (See Figure 1). The maximum pressure source, called the main header, is connected directly to the boiler and distributes steam at the maximum available pressure and temperature. The main header is supplied with sufficient capacity to carry the entire boiler(s) output without a significant loss of pressure to remote users.

Lower pressure steam is supplied by letdown (a reduction of pressure at a reducing station) from the main header. A typical one step letdown would be from 550 psig to 250 psig (See Table 1). This provides a source of steam to use at a pressure and temperature that is appropriate for each use in the process.

TABLE 1

STEAM PROPERTIES FOR TYPICAL PLANT SUPPLIES

Absolute Pressure psia	Temperature °F	Sensible (h_f) BTU/lb.	Latent (h_{fg}) BTU/lb.	Total (h_g) BTU/lb.	Typical Use
29.7	249.8	218.4	946.0	1,164.4	15 psig Low pressure steam used in tracing, jackets, tank heaters and etc.
30.7	251.7	220.3	944.8	1,165.1	
31.7	253.6	222.2	943.5	1,165.7	
32.7	255.4	224.0	942.4	1,166.4	
33.7	257.2	225.8	941.2	1,167.0	
34.7	258.8	227.5	940.1	1,167.6	
36.7	262.3	230.9	937.8	1,168.7	
264.7	406.1	381.7	821.2	1,202.9	250 psig Intermediate pressure steam for process use.
274.7	409.3	385.3	817.9	1,203.2	
284.7	412.5	388.8	814.8	1,203.6	
294.7	415.8	392.3	811.6	1,203.9	
304.7	418.8	395.7	808.5	1,204.5	
314.7	421.7	398.9	805.5	1,204.4	
554.7	477.8	462.0	742.8	1,204.8	550 psig High pressure steam for process use.
574.7	481.6	466.4	738.1	1,204.5	
594.7	485.2	470.7	733.5	1,204.2	
614.7	488.8	474.8	729.1	1,203.9	

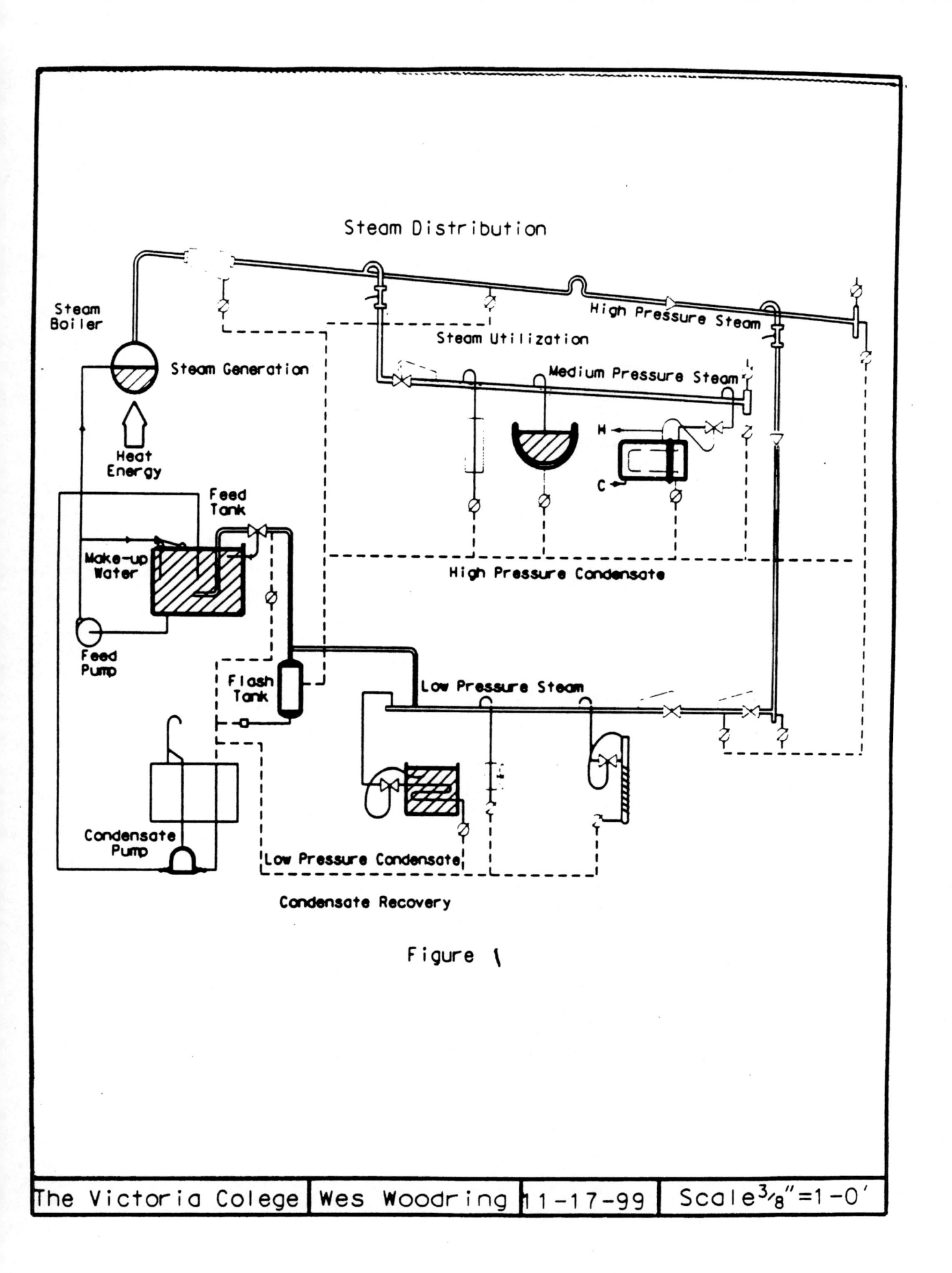

Figure 1

A common two step letdown would be from 550 psig to 250 psig to 15 psig. The principle to remember is that the important property of the steam supply is the header pressure, which establishes the maximum condensing temperature. The proper pressure gives good control of the temperature difference in heat exchange equipment.

Steam pressure is letdown to process needs:

1. Protect temperature sensitive process materials.

2. Reduce the design pressure of equipment.

3. Provide more sensitive temperature control.

One of the most important operating principles of any system operating at elevated temperature is allowance for expansion. A 2000 foot long steam line operating at 600 psig grows 3.2 feet when heated from an ambient temperature off 77°F to an operating temperature of 489°F. The expansion (growth in length and diameter) has to be provided for to prevent damage to piping, equipment, and supports. An expansion loop and movable feet are often the solution (See Figure).

All systems operating above ambient temperature lose heat to the atmosphere. The amount of heat lost is proportional to the temperature difference. When heat is lost from saturated steam, condensate is formed. In an eighteen inch line 2000 foot long being operated at 600 psig (489°F) the heat loss is 160,000 BTU/HR and results in condensate formation of 200 lbs./HR/100 foot of line, or a total 4000 lbs. of condensate.

Steam connections that feed the process are made to the top of headers for the purpose of leaving condensate and solids behind. Condensate connections are made to the bottom of headers to improve drainage into drip legs. Main headers are also installed

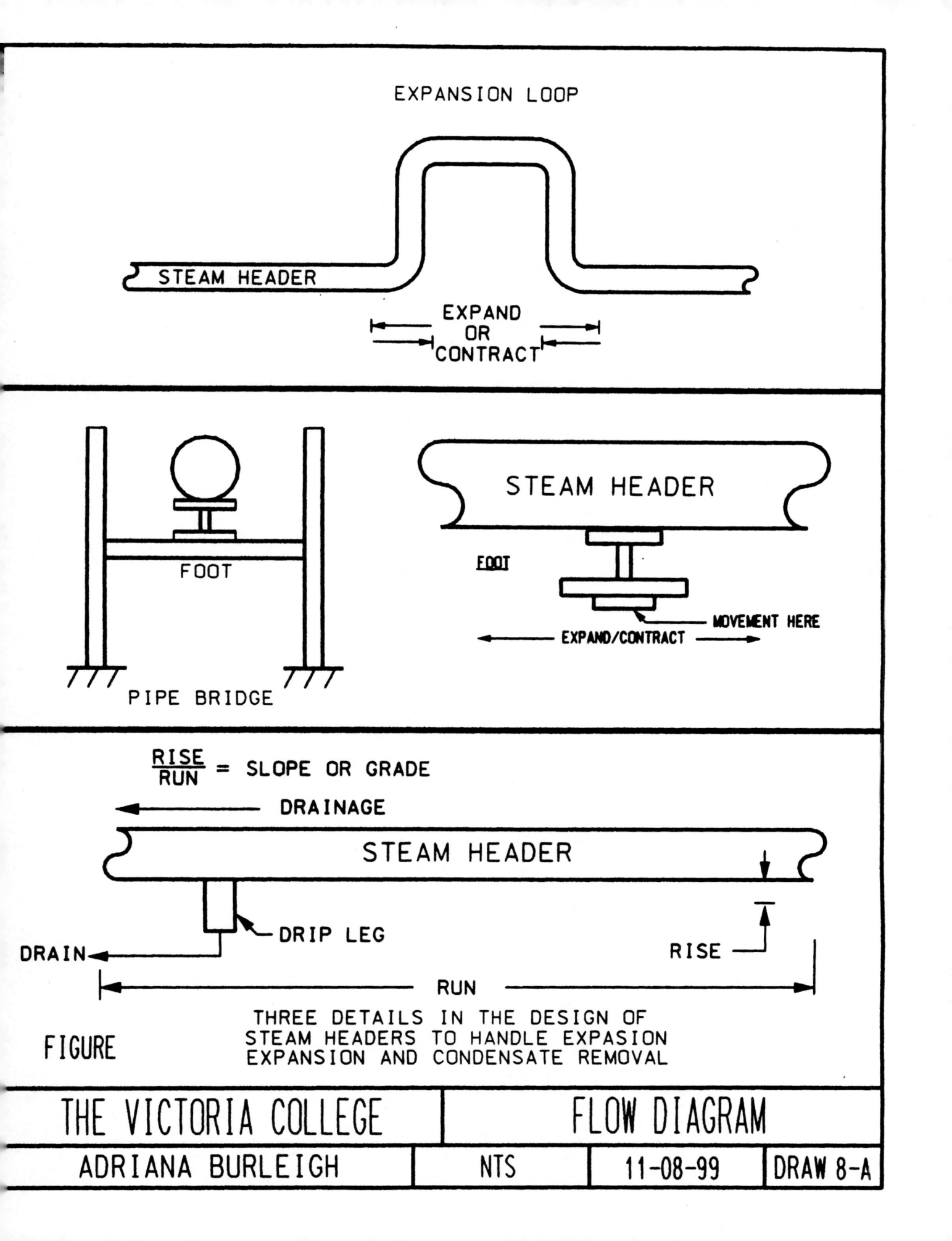

EXPANSION LOOP

STEAM HEADER

EXPAND OR CONTRACT

FOOT

PIPE BRIDGE

STEAM HEADER

FOOT

EXPAND/CONTRACT — MOVEMENT HERE

$\dfrac{RISE}{RUN}$ = SLOPE OR GRADE

DRAINAGE

STEAM HEADER

DRAIN ← DRIP LEG

RISE

RUN

FIGURE — THREE DETAILS IN THE DESIGN OF STEAM HEADERS TO HANDLE EXPASION EXPANSION AND CONDENSATE REMOVAL

THE VICTORIA COLLEGE	FLOW DIAGRAM		
ADRIANA BURLEIGH	NTS	11-08-99	DRAW 8-A

with a slight slope (rise over run) to facilitate good drainage in a controlled direction to the nearest drip leg.

Condensate is removed by steam traps at the condensing temperature. Condensate discharge pressure can be at any level from header pressure to the controlled pressure on the system. Back pressure on the trap is created by pressure drop in the condensate piping and the operating pressure of the flash tank when low pressure steam is being generated for process use. A trap is a mechanical device used to remove condensate from steam systems. A key element in any well-operated steam distribution system is the selection and maintenance of traps. Basically, a trap is a mechanical seal that passes liquid and retains vapor.

For good performance in traps and valves, as well as instruments in steam service, solids must be kept out. This is accomplished by use of strainers (filters) ahead of equipment that must be kept free of solids. Most solids are the result of scale formation, rust, or other corrosion products that break free from metal surfaces during temperature and pressure cycles.

Condensate piping is insulated only when it is necessary to protect personnel and for freeze protection. Normally, some loss of heat to the atmosphere is useful to prevent flashing.

When the pressure drop in a condensate header to the powerhouse is to high, the condensate is collected in a tank and pumped to the powerhouse. It is common practice to use this tank to produce low pressure steam for local use as tracing and other low pressure users. This is a very useful way to conserve energy (See Figure).

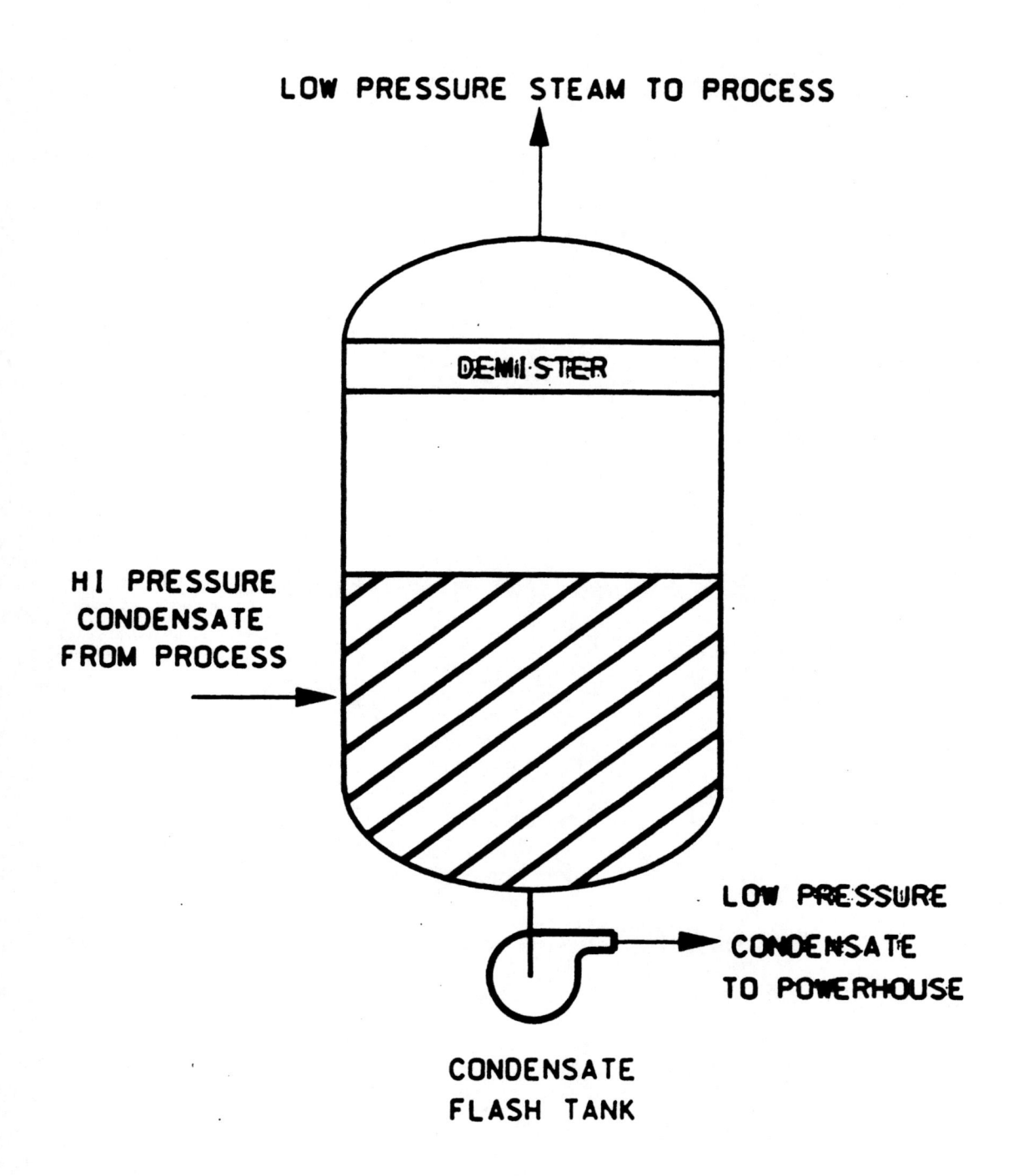

FIGURE:

FLOW DIAGRAM

Condensate and makeup water is combined in the deaerator where steam is used to strip out dissolved gases from the combined boiler field water.

STEAM TRAPS

High performance steam traps are indispensable for effective process control and for efficient use of the heat energy in steam. A trap can cause heat transfer performance if installed or maintained improperly. The objectives of the steam trap are simple: discharge steam condensate at its boiling point (constant pressure) from a steam user while maintaining a vapor lock. This creates a constant condensing temperature, which is necessary for effective heat transfer control through use of mean temperature difference.

Steam trap operation is based on several thermodynamic and thermostatic properties, which include controlled thermal expansion and contraction of liquids, and metals as steam condensate cools.

Balanced pressure thermostatic trap (See Figure A)

The key to the balanced pressure thermostatic trap is the bellows. The bellows is filled with a special liquid and is permanently sealed. The liquid has temperature vapor pressure properties close to but below water.

When cold, the bellows are fully contracted and any material fed to the trap is freely discharged through the open valve.

As condensate heats the trap, the bellows expands due to the buildup of internal vapor pressure inside the bellows, which exceeds the pressure in the trap body. The bellows expand which closes the valve and allows the trap to fill with condensate.

As the trap loses heat to the atmosphere, the vapor in the bellows condenses, the pressure balances, the bellows contracts and the valve opens to discharge condensate.

Fresh hot condensate enters the trap; the bellows expand which closes the valve. The contraction and expansion cycle repeats as often as needed to discharge condensate.

One potential problem is that the trap capacity is dependent upon heat loss to the atmosphere, which must be monitored.

Liquid Expansion Thermostatic Trap (See Figure B)

This trap is similar, in operation to the balanced pressure unit. However, no vapor is formed from the cylinder liquid, which moves the plunger with simple thermal expansion.

When hot condensate enters the trap, the liquid in the cylinder expands which drives the plunger and closes the valve. As the condensate cools due to heat loss to the atmosphere the liquid in the cylinder contracts, the plunger retracts and the valve opens. The condensate is discharged. The plunger being spring loaded which insures positive movement.

Float and Thermostatic Trap (See Figure C)

The advantage of the float trap is continuous discharge of condensate as it enters the trap. The liquid level is maintained above the discharge port to provide a positive liquid seal.

The trap is equipped with a thermostatic air valve, which can be adjusted to control trap pressure.

Inverted Bucket Trap (See Figure D)

The bucket trap works on the density difference between steam and condensate. The inverted bucket opens and closes the discharge valve as it rises and sinks in condensate. This is the result of being either vapor or liquid filled.

The trap is empty on startup so is in a fully down position as air is discharged. As condensate fills the trap, steam is generated in the bucket and it rises to close the discharge valve. As the trap cools due to heat loss, the steam condenses the bucket fills with condensate, and sinks. The valve opens and condensate discharges. There is a small vent at the top of the bucket to prevent buildup of non-condensables such as air.

Thermodynamic Steam Trap (See Figure E)

This trap has only one moving part, which opens or seals the condensate port.

As condensate pressure builds under the disc, the trap cools which lowers the pressure on the disc top and it opens allowing condensate to enter the trap. The hot condensate entering the trap increases the pressure on the trap top, lowers the pressure on the bottom and the disc falls which closes the trap. As the trap cools, the cycle repeats.

General

Traps depend upon heat loss to the atmosphere to cycle the operating temperature and pressure, which activates a useful physical change inside the trap. That opens and closes a discharge port.

Trap arrangement is an important design factor. Pressure losses or gains due to pipe size; fittings and elevation can all create pressure drop, which can prevent the trap from functioning properly.

Also, trap elevation increases or reduces condensate head, which will effect trap performance.

Any proposed trap installation that changes operating mode, size, or location must be carefully designed by a knowledgeable trap expert.

FIGURE
TYPICAL STEAM TRAPS

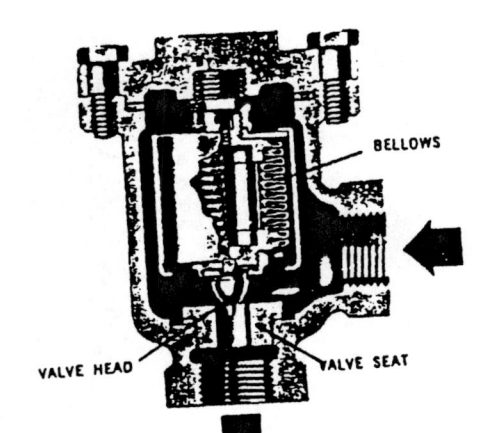

BALANCED PRESSURE THERMOSTATIC

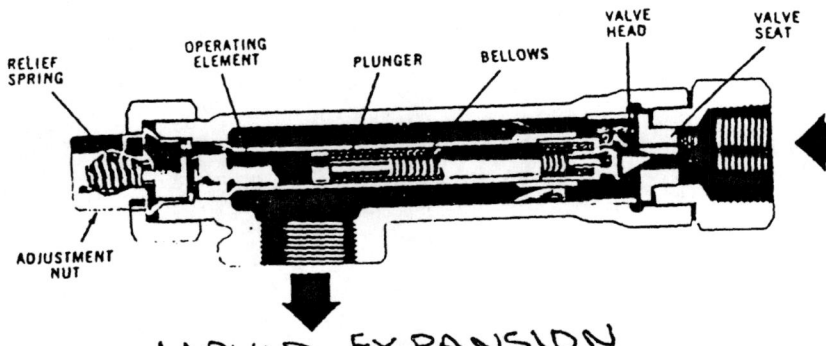

LIQUID EXPANSION THERMOSTATIC

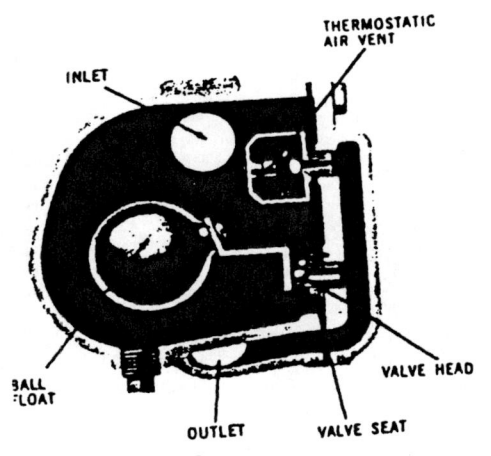

FLOAT AND THERMOSTATIC

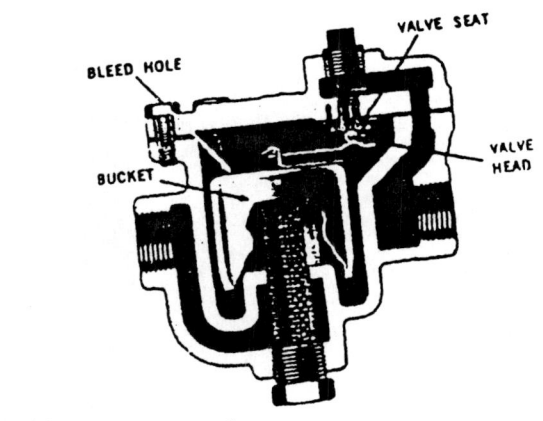

INVERTED BUCKET

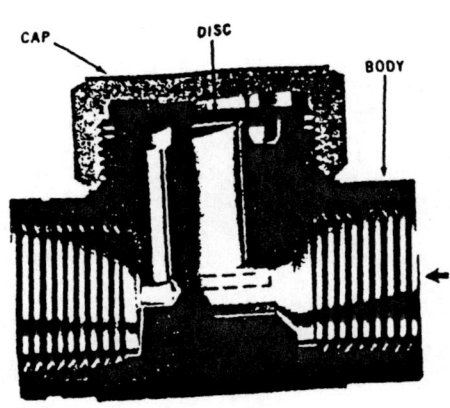

THERMODYNAMIC

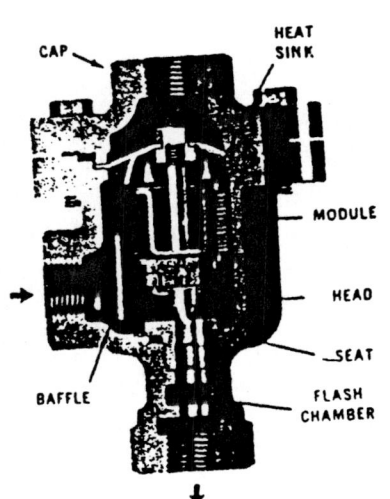

THERMOMATIC

TERMS AND SYMBOLS

Header – A main steam distribution pipe usually located on a pipe bridge.

Pipe Bridge – An above ground steel frame that supports piping between buildings and between processes.

Linear Expansion – The growth or elongation in length of a material due to an increase in temperature. The coefficient of expansion is usually expressed as FT/FT/°C (feet of growth per foot of length per °C).

Expansion Loop – A u-shaped section in a pipeline designed to safely flex with temperature change, which relieves thermal stress due to elongation.

Trap – An equipment piece designed to discharge condensate (water) and maintains a vapor seal at constant pressure.

Reducing or Letdown Station – When pressure is deliberately reduced to a controlled level across a device such as a valve, it is called letdown or reducing station. Several levels of steam pressure are often produced in one station.

Condensate Collection Pipe or Drip Leg – All steam systems lose heat to the atmosphere, which produces condensate. The condensate is collected in and discharged from a collection pipe called a drip leg.

Flash Tank – Condensate is saturated water, which means that a reduction in pressure generates vapor in the form of steam. To reduce pressure drop, prevent pump cavatation and to provide a low pressure steam source, the condensate pressure is reduced in a vessel called a flash tank.

Thermodynamic – Refers to the physics of energy forms due to an increasing or decrease in heat energy.

Thermostatic – Refers to a mechanical movement or direction due to a temperature gradient.

Saturated Water – Saturated Steam – At constant temperature and pressure, water is in equilibrium with its vapor pressure. An increase in temperature or decrease in pressure produces more steam (vaporization). A decrease in temperature or an increase in pressure produces more water (condensation).

SAFETY

Any fluid under pressure is potentially hazardous. The human eye is especially vulnerable. At the elevated temperatures in steam and condensate lines, the burn hazard is always present.

It is important to open valves in the condensate system with care. Condensate is saturated water at its boiling point and flashes to steam when the pressure is released. Drains and vents should be opened with care with the hot condensate directed away from any part of the body.

Initial heat-up of steam systems is carried out at a careful rate to allow even expansion throughout the main headers and supports. Since this may well be an infrequent occurrence, review of the procedures is good safety practice.

One danger in an undrained or improperly drained steam main lies in the possibility of water hammer (See Figure). Water hammer is the impact caused by sudden stopping of a rapidly moving slug of water.

Unless condensate is removed from low points in mains, it gradually accumulates until the fast flowing steam causes ripples on the surface. Eventually, the condensate so restricts steam flow that a slug of condensate is carried down the main by the steam. The slug of water travels at the speed of the steam (which may be in excess of 100 miles/hour) until some obstruction is reached. The obstruction may be a reducing valve, temperature regulator, a steam trap, a change in direction of the main, or simply the end of the line. In any case, the slug of water is suddenly stopped with an often released of kinetic energy to potential energy. Property damage by water hammer is serious. Potential injury is prevented by careful consideration of steam main drainage.

CHAPTER 6

REFRIGERATION

INTRODUCTION

Refrigeration is the simplest term used to describe the production of cold conditions below atmospheric temperature. Refrigeration is put to work to cool food, homes, cars, and businesses. Everyday living has become unthinkable without it.

The same technology used in the private section to produce cold is used in Industrial Processes to perform work at reduced temperatures where that same work would be impractical at normal temperatures. The process most commonly used is to cause a phase change in a heat-abstracting medium. This could be the melting of ice or the vaporization of a refrigerant like Freon.

A refrigerant in its broadest sense may be defined as any material which is used for abstracting heat. In a narrower but more commonly used sense, the term refers only to those materials. which are used in mechanical refrigeration.

Mechanical refrigeration includes those processes in which the refrigerant is recovered and re-circulated, as distinguished from those in which the spent refrigerant is wasted (ice refrigeration, ice cream freezing by salt-ice mixtures, cold ground waters in springhouses.)

THE PHASE CHANGE (Diagram 1)

The Phase change (Figure 1) is a key physical principle used in distillation, steam generation, refrigeration, and other unit operations.

The conversion of a substance from solid to liquid, to vapor by the addition of heat is referred to as phase changes. What is useful about a phase change is that when the

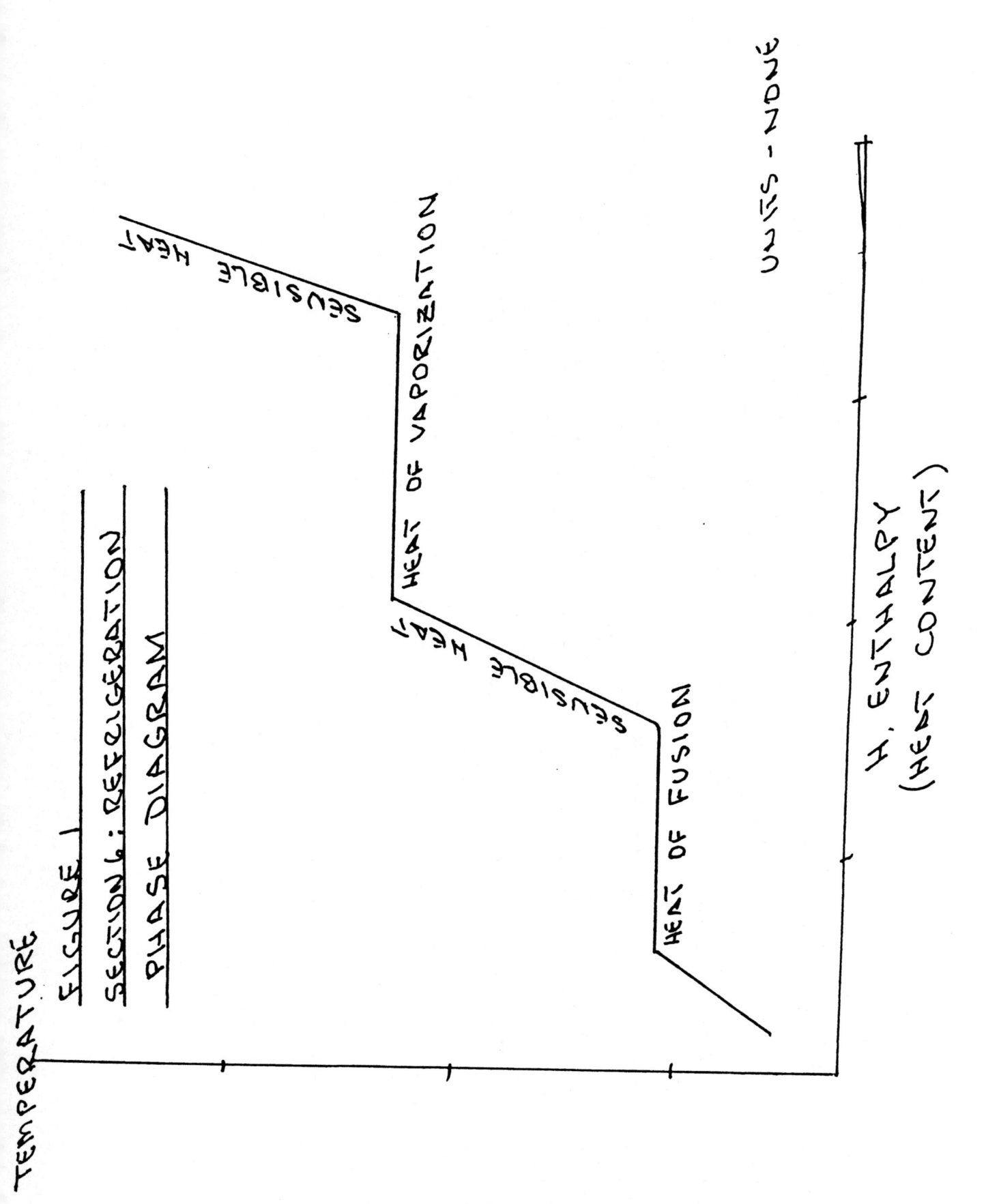

Figure 1
Section 6: Refrigeration
Phase Diagram

Temperature vs *H, Enthalpy (Heat Content)*

Regions (low H to high H): Heat of Fusion, Sensible Heat, Heat of Vaporization, Sensible Heat.

Units – none

same amount of heat energy is removed and the vapor becomes liquid. Additional heat removal and the liquid become solid. These are also phase changes.

The second key point is that the phase changes from solid to liquid and liquid to vapor or the reverse take place at a constant temperature and pressure until the phase change is complete. Stated another way, when a solid like water (ice) begins to melt it stays at the melting temperature until all the ice is converted to liquid. The ice is taking on latent heat energy called the heat of fusion. The vapor pressure of a liquid reaches the surrounding pressure at the boiling point temperature. A liquid vaporizes at the boiling point temperature under constant pressure and remains at that temperature until all of the liquid is converted to vapor. The latent heat energy required to convert the liquid to vapor is called the heat of vaporization (Hv). Latent heat is very large compared to sensible heat.

At this point, it has been demonstrated that a liquid can be converted (evaporated) by adding latent heat (heat of vaporization) at constant temperature and pressure. It has also been shown that a gas goes back to its liquid state (condensation) by removing the same amount of latent heat (heat of condensation) at constant temperature and pressure. When substance is repeatedly vaporized and condensed with the addition and removal of a fixed amount of heat energy from a closed system, the process is said to be reversible.

<u>What is also true is that the same heat energy can be moved back and forth between liquid and vapor at a higher or lower temperature by increasing or decreasing the pressure.</u>

VAPOR COMPRESSION CYCLE

The vapor compression cycle is diagrammed on Figure 2. The objective is to use the latent heats of vaporization (Hv) and condensation (Hc) to remove heat energy at low temperature from a process stream and dispose of that same quantity of heat energy at a more convenient pressure and temperature.

At point A on Figure 2, the liquid refrigerant has been converted to vapor by adding the latent heat of vaporization starting at point D. The conversion is usually carried out in an equipment piece called an evaporator. And the latent heat is supplied by a process stream that is cooled (refrigerated).

At point A, the enthalpy (heat content) of the refrigerant is increased and is equal to the heat of vaporization, which can vary with the refrigerant in use.

The refrigerant vapor is compressed from pressure A to pressure B. In the compression system a compressor is used, which may have either a positive displacement mechanism (reciprocating) or an impeller (centrifugal compressor). The thermodynamic cycle is the same for both types of compression. This will add a small quantity of heat energy (heat of compression) which increases temperature. The increase in pressure increases the boiling point, which becomes the new condensing temperature.

The next step is to condense the vapor to a liquid by removal of the latent heat at the higher pressure and temperature. The heat of condensation is removed between B and C and at C the refrigerant vapor is all converted but at a higher pressure. The heat is conveniently removed in cooling water because of the increase in the boiling point. The liquid is stored at the higher temperature and pressure in a reservoir (tank) or in an accumulator, for smaller units until fed to the evaporator through a letdown valve between C and D.

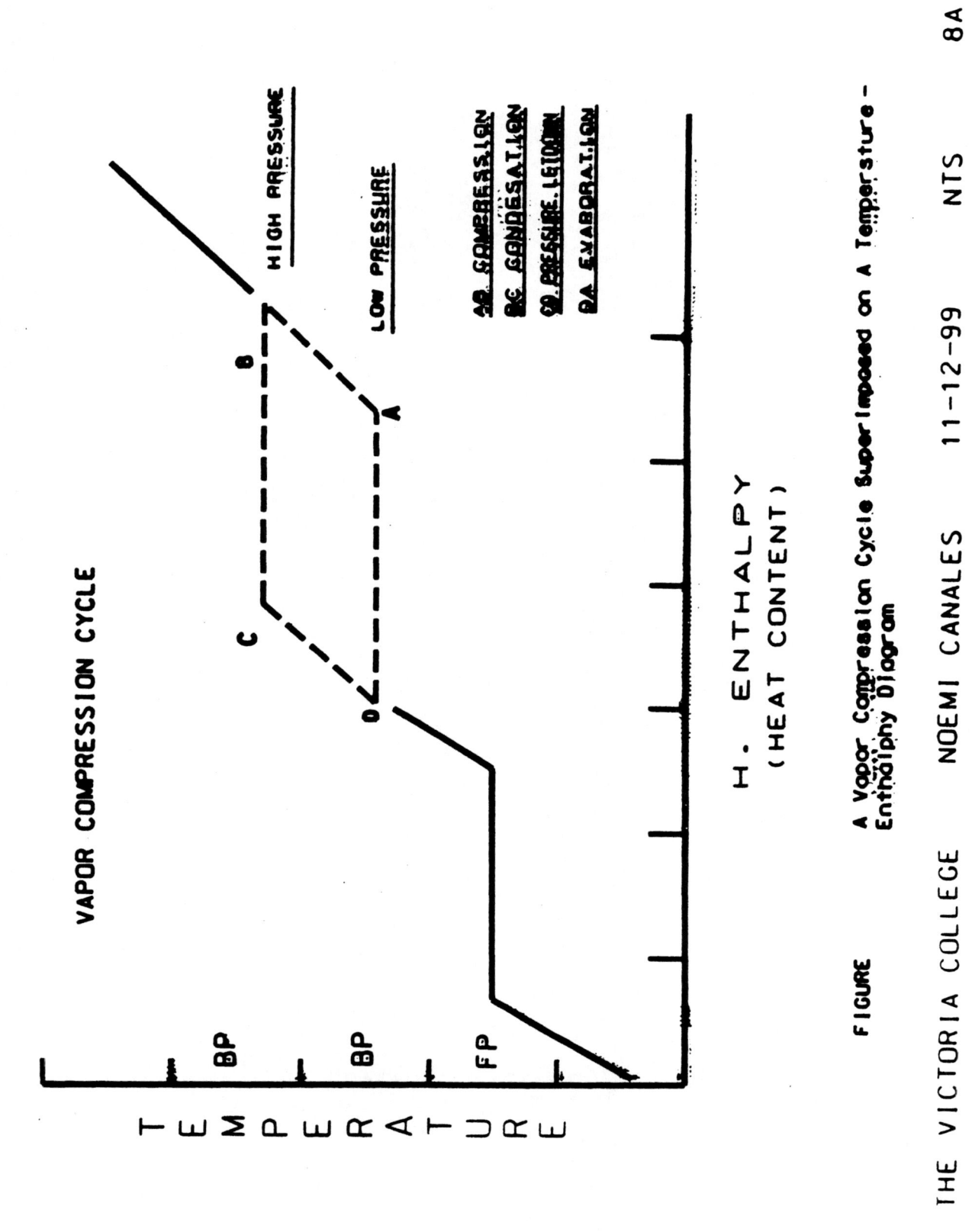

VAPOR COMPRESSION CYCLE

HIGH PRESSURE

LOW PRESSURE

AB COMPRESSION
BC CONDENSATION
CD PRESSURE LETDOWN
DA EVAPORATION

Y-axis: TEMPERATURE (BP, BP, FP marked)
X-axis: H. ENTHALPY (HEAT CONTENT)

FIGURE: A Vapor Compression Cycle Superimposed on A Temperature – Enthalphy Diagram

THE VICTORIA COLLEGE   NOEMI CANALES   11-12-99   NTS   8A

The refrigerant liquid will convert to vapor at a much lower temperature with the reduction in pressure, which lowers the boiling point. The latent heat of vaporization is added as shown between D and A to complete the phase change at constant temperature. The heat of vaporization comes from the stream being cooled and the cycle is complete.

The cycle can be continuously repeated in a closed system with exact amounts of heat energy transferred out of a process and into a coolant like water using a material called a refrigerant. The overall process is called the reversed ranking cycle.

Major Components in a Vapor Compression Cycle (See Figure 3) [2,1]

1. In a reservoir or receiver, liquid refrigerant is stored under high pressure. The temperature in the reservoir is at the boiling point of the liquid at the reservoir pressure. Temperature and pressure are controlled by bleeding off vapor formed from outside heat.

2. The liquid pressure is reduced across an expansion valve (pressure letdown). At the reduced pressure, the liquid refrigerant will vaporize at the lower boiling point if the latent heat of vaporization is provided. The low temperature point is established as the boiling point of the refrigerant at the reduced pressure.

3. The vaporizer (sometimes called an expander or evaporator) can be a coil, heat exchanger, or perhaps a radiator.

The heat of vaporization is provided a process stream that is cooled toward the boiling point of the refrigerant. The temperature difference between the process and the boiling refrigerant is the ΔT required to transfer heat energy. This quantity of heat energy transferred is the heat load on the refrigeration unit.

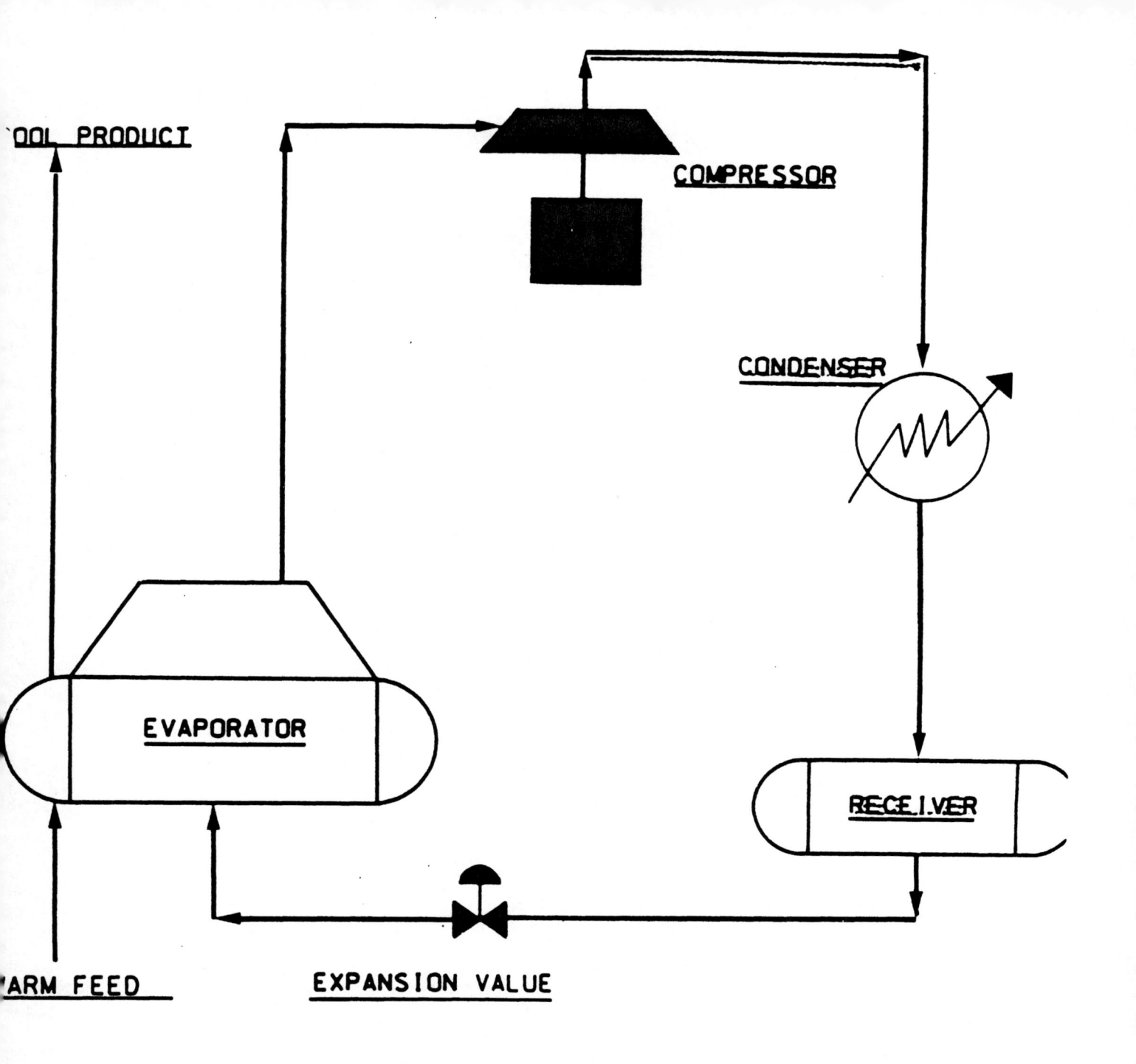

FIGURE 12.1 REFRIGERATION
VAPOR COMPRESSION COMPONENTS

4. The refrigerant is now converted to a vapor and contains its latent heat of vaporization. The heat that was transferred into the refrigerant from the process is of no practical use and a method to discard it is needed at a reasonable operating temperature and pressure. This is accomplished in two steps. The vapor is compressed to increase the condensation temperature (boiling point). This establishes the desired condensing temperature. The heat of compression (work on the vapor) increases the temperature.

   The higher pressure needed to condense the vapor at a higher temperature is now set which allows the latent heat of vaporization and the heat of compression to be transferred into a convenient disposal medium such as cooling water. The overall effect is to transfer heat from the process at a low temperature to provide latent heat in the evaporator and transfer that same latent heat to the cooling water at a convenient temperature in a condenser.

5. Energy Balance

   $$\text{Efficiency} = \frac{H_b - H_e}{H_a - H_b}$$

   H_b = Heat content exit vaporizer, BTU/lb.
   H_a = Heat content exit compressor, BTU/lb.
   H_e = Heat content into vaporizer, BTU/lb.

   Useful work = $H_b - H_e$
   Work of compression = $H_a - H_b$

6. The liquid refrigerant is stored in the reservoir and ready for reuse in the evaporator. The cycle is complete.

THE ABSORPTION CYCLE

The absorption cycle is one of the oldest and most practical of the refrigeration cycles (See Figure). One major advantage is the use of an inexpensive material such as ammonia as the refrigerant and water as an absorbent. A second advantage is the elimination of expensive compression facilities in order to condense the ammonia at a practical operating temperature.

The absorption system differs essentially from the compression system in requiring no positive work input to produce cold through generally requiring some accessory power. Circulation is effected by absorption of the refrigerant in an appropriate liquid are regenerated by heat in another part of the refrigerating system.

FIGURE

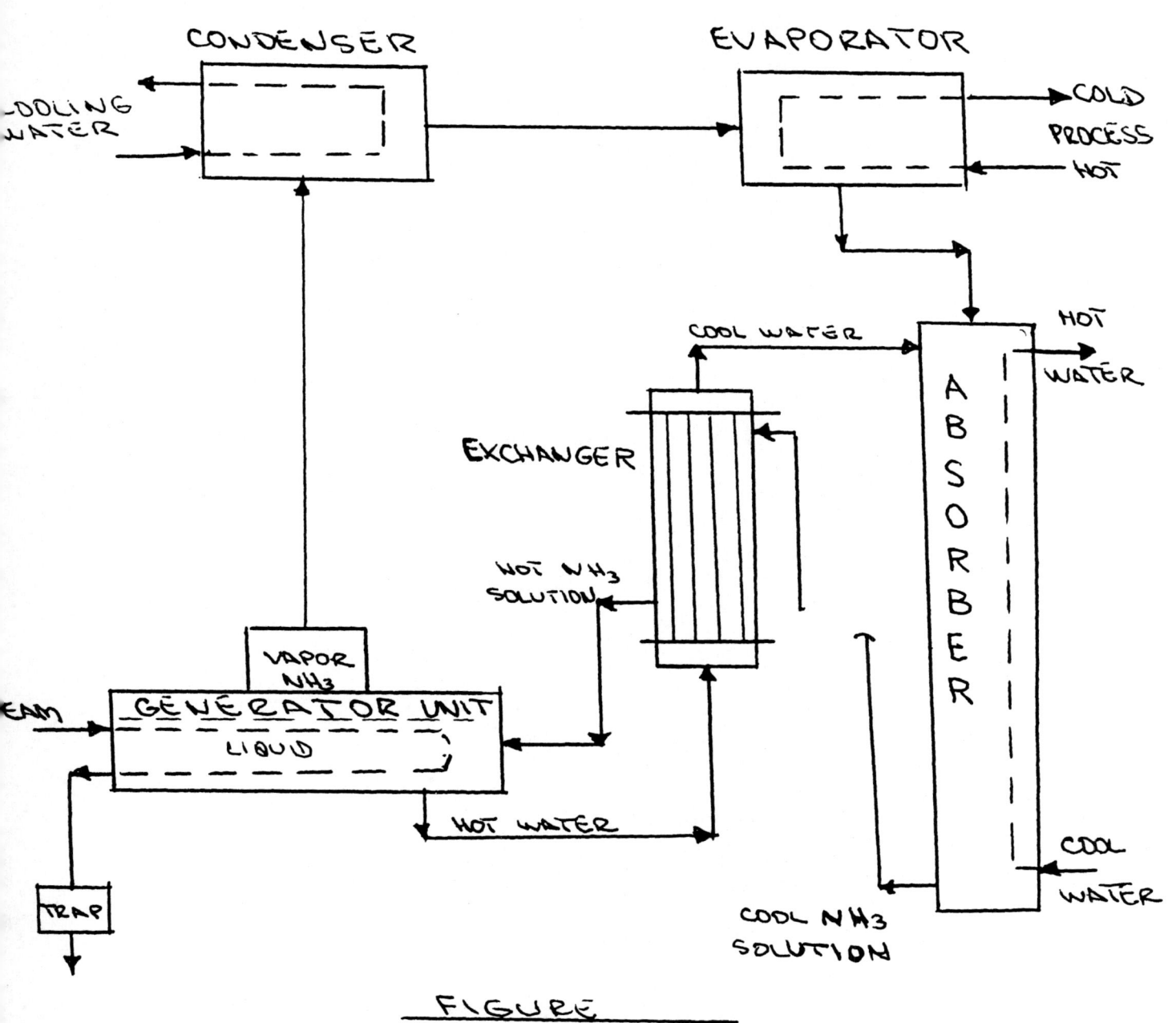

REFRIGERATION

AMMONIA (NH₃) ABSORPTION SYSTEM

CASCADE REFRIGERATION

There are refrigeration needs such as liquification of natural gas where a single vapor compression cycle cannot span the overall temperature range needed to remove latent heat at a very low temperature and discard it at close to atmospheric temperature. One answer is to install two vapor compression cycles in series. This in effect creates cooling and heating in two practical steps.

FIGURE
REFRIGERATION CASCADE
LIQUIFACTION OF NATURAL GAS

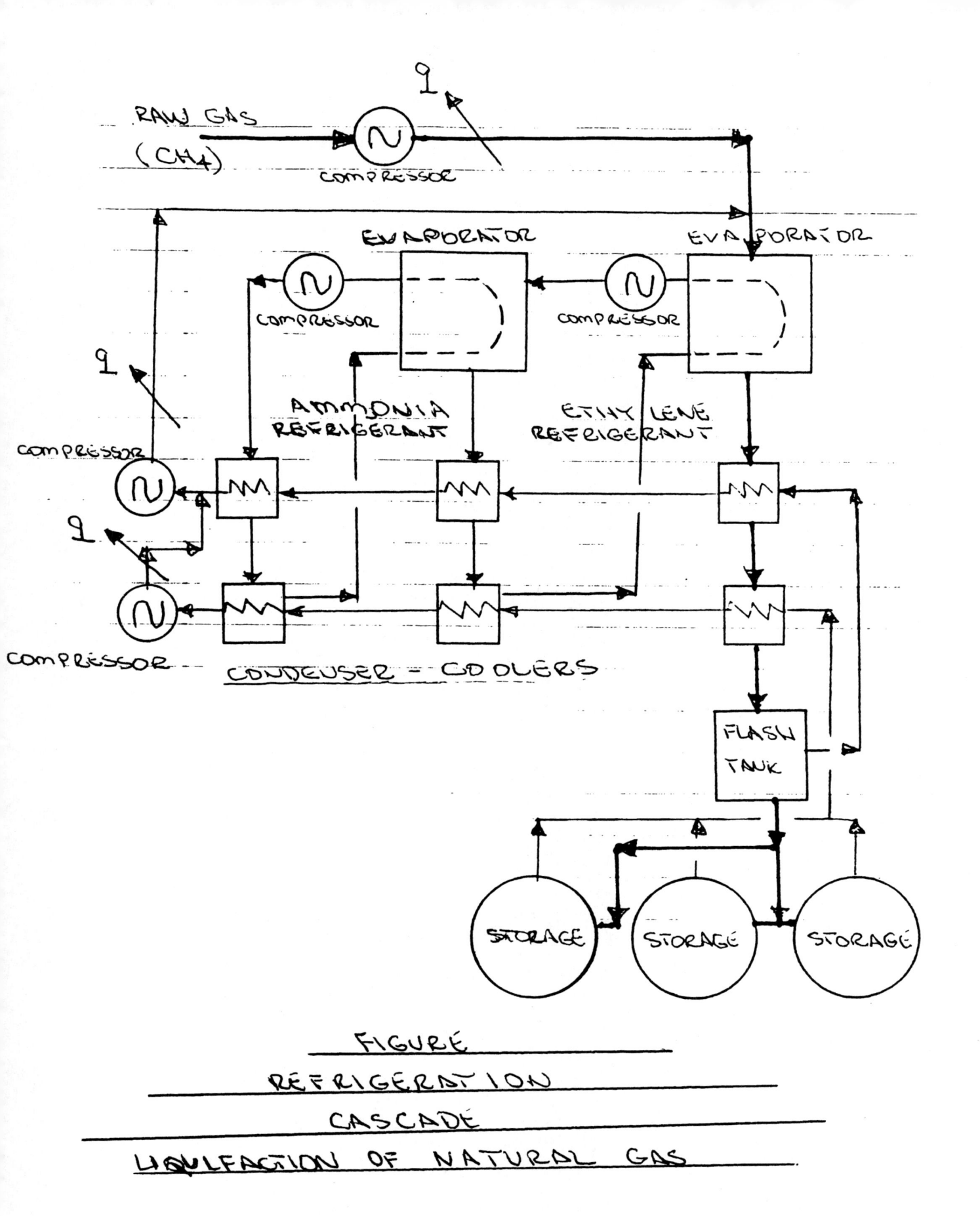

THE HEAT PUMP

A machine called a heat pump that is capable of either adding or removing heat to and from two locations has been developed for use in moderate climates.

In the summer, the unit performs as an air conditioner. Heat is transferred from the air being cooled into the refrigerant to provide latent heat.

In the winter, heat is transferred into the air being heated by providing the heat of condensation to the refrigerant vapor.

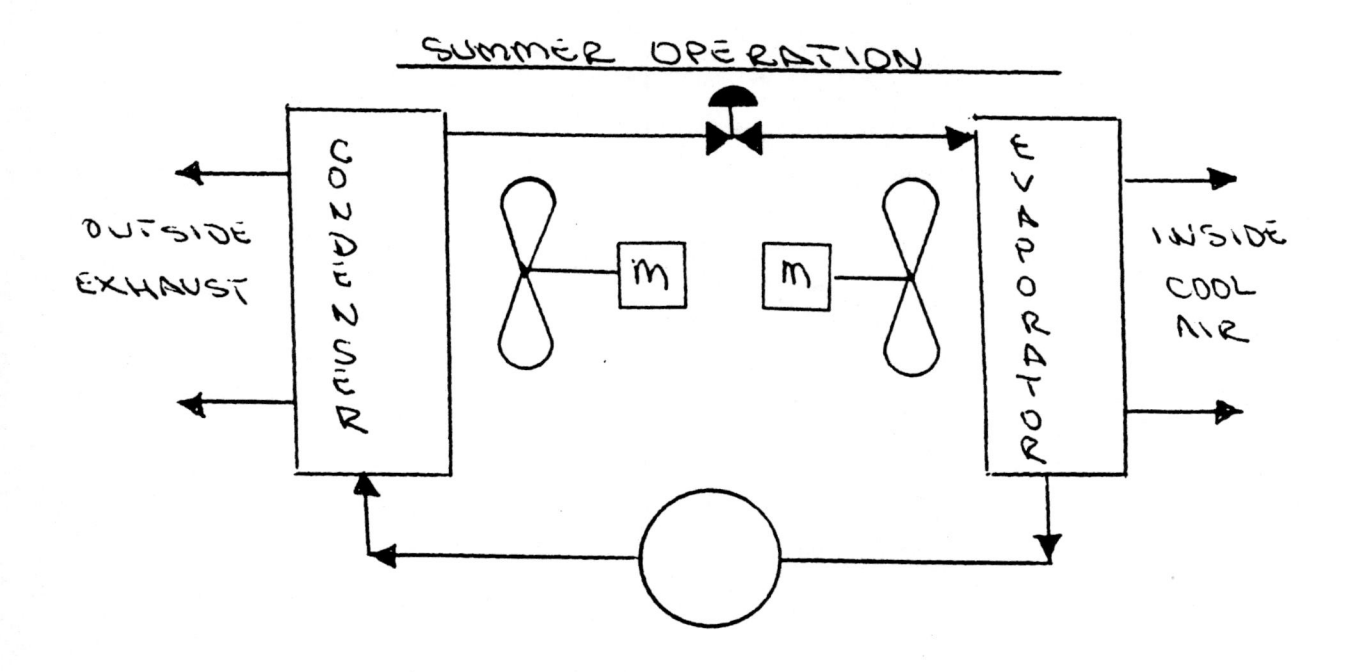

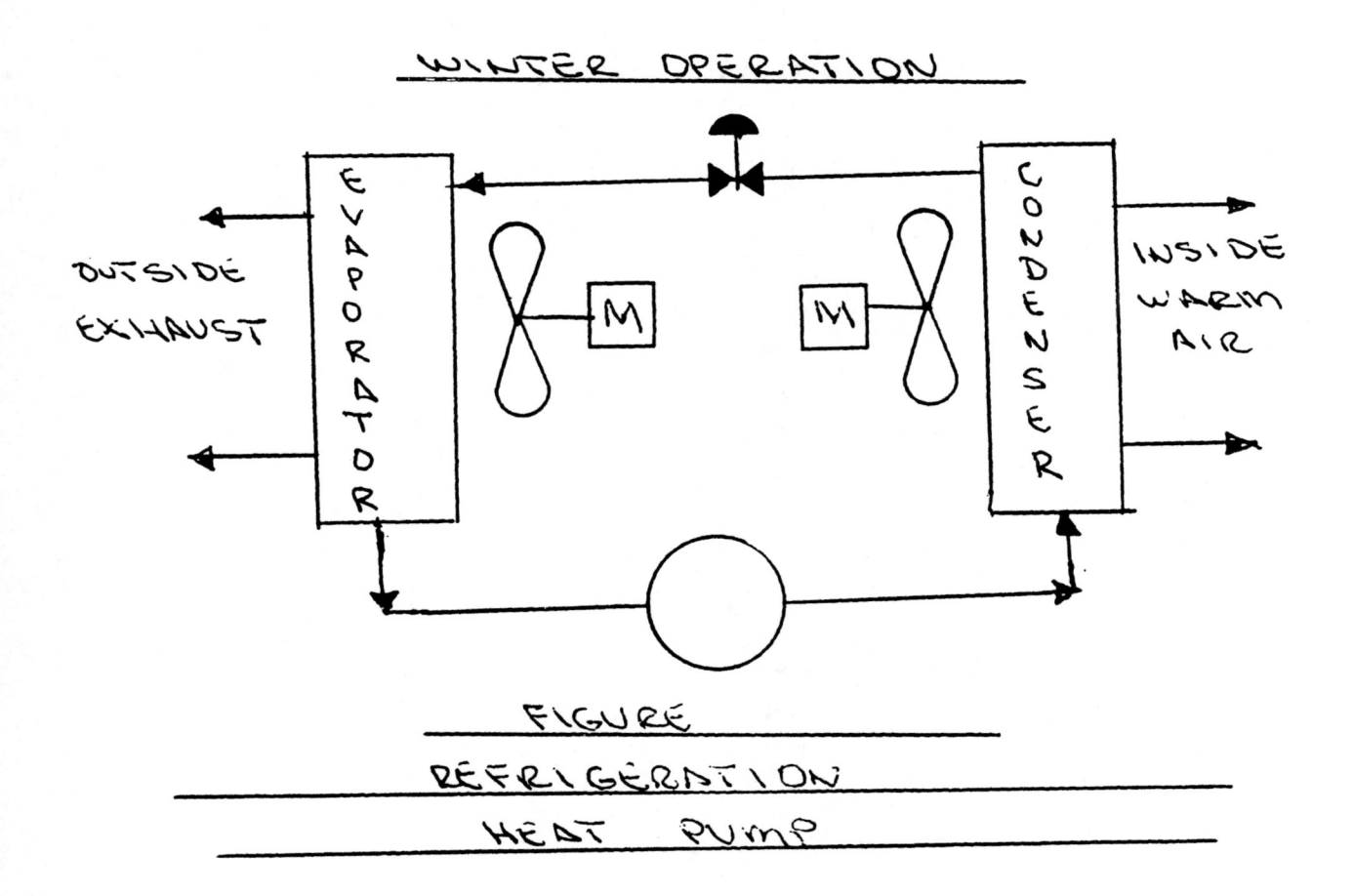

FIGURE
REFRIGERATION
HEAT PUMP

UNITS OF REFRIGERATION

The unit of refrigeration in the United States is the standard ton of 288,000 BTU, which is very nearly equal to the heat of fusion of 2000 lb. of ice at 32°F. The standard commercial ton of refrigeration is at the rate of 200 BTU/min., 12,000 BTU/hr., or 288,000 BTU/24 hr. Note that the standard ton has the dimensions of heat, while the standard commercial ton has the dimensions of heat divided by time.

TERMS AND SYMBOLS

Refrigerant – A chemical that evaporates and condenses at a convenient temperature and pressure that provides the low temperature needed by the process when evaporated and a high temperature needed for efficient disposal of the heat energy when condensed.

Vapor Compression Cycle – Utilizes the latent heat of vaporization to remove heat and in turn uses the latent heat of condensation to dispose of that same heat.

Commercial Standard Ton – The term used to define the capacity of a refrigeration unit. One ton is a heat removal capability of 12,000 BTU/hr.

Brine – A salt solution that is chilled in a refrigeration unit for use as a process coolant in remote locations.

Evaporator – An equipment piece used to convert a liquid to a vapor at a convenient temperature and pressure using an external source of heat energy input.

Condenser – An equipment piece used to convert a vapor to a liquid at a convenient temperature and pressure using an external source of heat energy removal.

Adiabatic Compression – Takes place when no external heat energy enters or internal heat energy leaves the compression system.

Heat of Compression – The work done to increase pressure on the gas and results in an increase in temperature at adiabatic conditions. Heat of compression adds to sensible heat of the system.

Package or Turnkey Unit – Designed and installed with a guarantee of performance.

CHAPTER 7

DISTILLATION

INTRODUCTION

Distillation is a widely used unit operation that can be used to separate two or more miscible components in a liquid mixture. The equipment is usually large with considerable amount of materials and energy involved.

Distillation performance depends upon the control of a large number of interdependent process variables to maintain steady state conditions at a very high level. Understanding how the process variables interact is fundamental to understanding how a distillation process is controlled in order to achieve high separation performance.

Distillation facilities are not uniquely complicated and the process technology is well defined. However, there are several process conditions that must exist before distillation can be considered the best choice of separation techniques.

First, each component in the mixture to be separated must have a vapor pressure distinctly different at the proposed operating temperature. A significant difference in vapor pressure also implies a distinct difference in boiling point for each component.

Another requirement for effective separation by distillation is that each component to be separated must be present in the original mixture in a reasonable concentration.

This component must be chemically inert and thermally stable at all distillation conditions. Decomposition at elevated temperatures is a common problem. A practical solution is to distill under vacuum as a means of lowering the boiling points.

Finally, the components should remain in the desired liquid or vapor state with no solids allowed to form at the operating temperature and concentrations. All the above variables will be fully explained.

Some distillations require too many actual stages to be practically housed in a single column. Therefore, distillation operations called trains contain two columns or more. Each column is given a descriptive name.

<u>Topper</u> is a column that removes low boilers.

<u>Refiner</u> is a column that produces the desired product.

<u>Low boiler</u> is a column that concentrates or refines low boilers.

<u>High boiler</u> is a column that concentrates or refines high boilers.

<u>Concentrator or Concentrating Still</u> is a column used to remove a diluent like water.

PHYSICAL BEHAVIOR OF MIXTURES
IN THE DISTILLATION PROCESS

The driving force in the distillation process is the vapor pressure difference between the individual components in a mixture at the boiling point as vaporization is taking place.

The general principle of vaporization is the same regardless of whether it takes place in a lab flask or in a huge distillation column. To produce vapor, a liquid is boiled. For a pure liquid to boil, its saturated vapor pressure must equal the system pressure immediately above the surface of that liquid. The vapor pressure of a liquid indicates the equilibrium condition between molecules in the liquid phase and molecules in the vapor phase. That is, at the given temperature in a closed container the number of molecules escaping from the liquid equals the number of molecules that return to the liquid. The higher the temperature the greater the number of molecules that escape which increases the vapor pressure. Until the boiling point of the liquid is reached, the number of molecules that escape the liquid will equal the number of molecules that return to the liquid. When a liquid begins to boil, a greater number of molecules escape from the liquid that the number returning to the liquid. This process called vaporization, continues at constant temperature and pressure for pure liquids until the phase change of the liquid to vapor is complete.

A mixture has physical properties that fall between the extremes of the pure components (See Figure 1 and Table 1).

TABLE 1

Physical properties of a mixture effected by concentration.

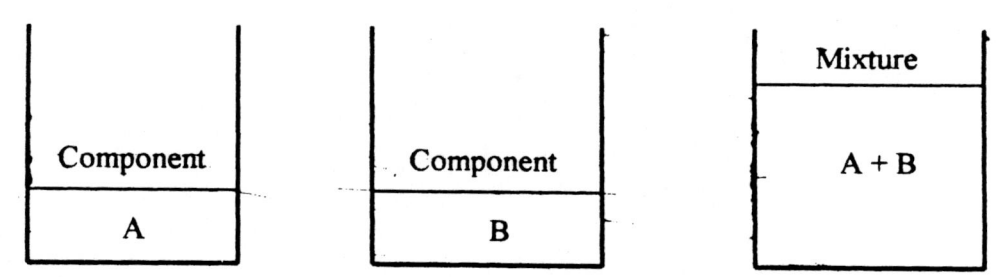

1. A and B are miscible fluids.

2. When mixed (A + B), A and B share their individual properties and give the mixture of A + B an intermediate set of properties.

3. The properties of A and B as well as the properties of A + B of the mixture are determined by the concentrations of A and B measured in either weight or molecular concentrations.

Liquid Molecular Concentrations	Liquid Weight Concentrations
Vapor pressure	Density
Boiling point	Heat capacity
Vapor volume	Heat of vaporization
Vapor molecular composition	Heat of condensation

FIGURE 1

Vapor Pressure and Boiling Point

Of a Mixture

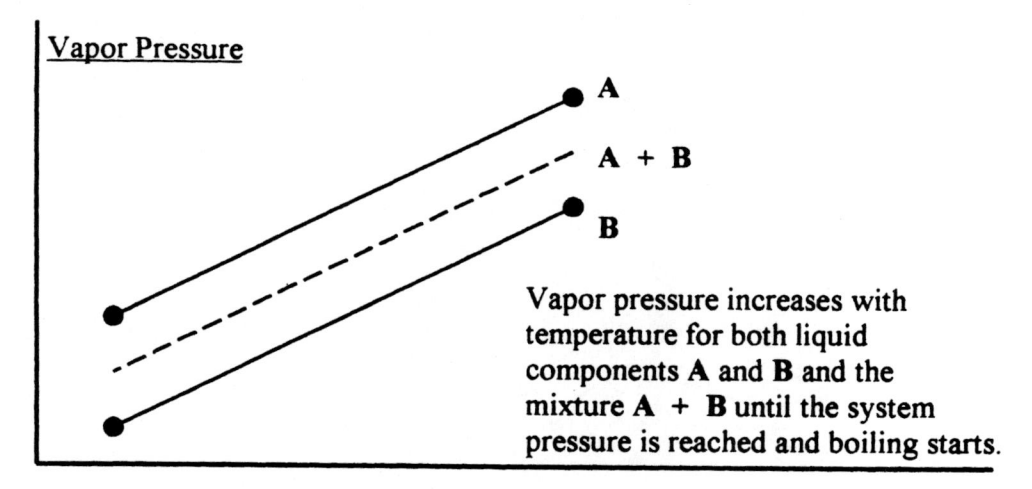

Vapor Pressure

- A
- A + B
- B

Vapor pressure increases with temperature for both liquid components **A** and **B** and the mixture **A + B** until the system pressure is reached and boiling starts.

Temperature

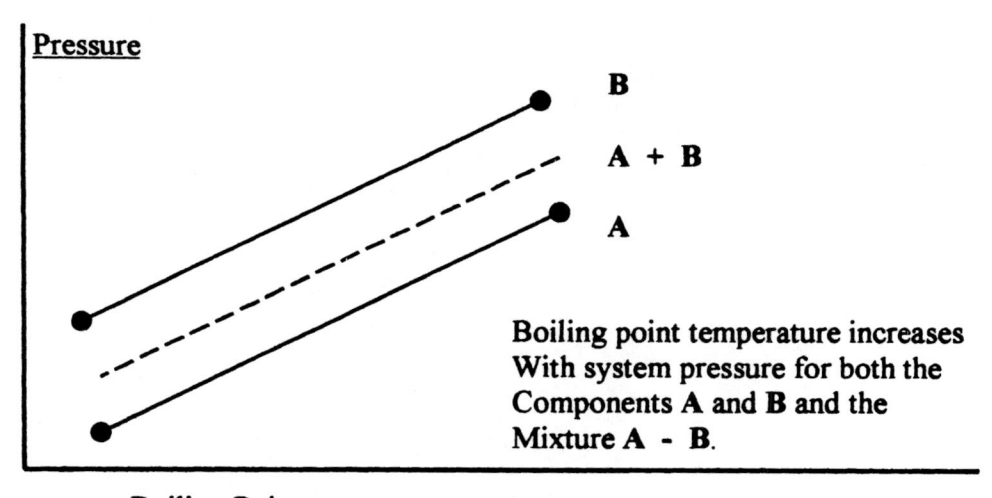

Pressure

- B
- A + B
- A

Boiling point temperature increases With system pressure for both the Components **A** and **B** and the Mixture **A - B**.

Boiling Point

Distillation works as a separation process through use of vapor pressure difference and molecular concentration differences of miscible components in a mixture. The relation ship is stated in Raoult's Law.

Raoult's Law states that:

The partial pressure of a component in the vapor in equilibrium and contact with a liquid mixture is equal to the vapor pressure of the pure component in the liquid mixture times its <u>molecular concentration</u> in the liquid mixture.

For a mixture that contains miscible components **A** and **B**, the equations can be written as:

$$P_A = (V_P \text{ of } A)(C_A)$$

<u>P_A</u> is the partial pressure of component **A** in the vapor in contact with liquid **A** and **B**.

<u>V_P of **A**</u> is the vapor pressure of pure **A**.

<u>C_A</u> is the molecular concentration of component **A** in the liquid **A** and **B**.

$$P_B = (V_P \text{ of } B)(C_B)$$

<u>P_B</u> is the partial pressure of component **B** in the vapor in contact with liquid **A** and **B**.

<u>V_P of **B**</u> is the vapor pressure of pure **B**.

<u>C_B</u> is the molecular concentration of component **B** in the liquid **A** and **B**.

The distillation process has many important variables that must be considered for stable operation.

For this study, three are going to be fixed.

1. The concentrations of components **A** and **B** in the feed stream are assumed to be constant.
2. The separation by distillation of **A** and **B** is assumed to be ideal which means separation fully obeys Raoult's Law.

3. The vapor components **A** and **B** obey the gas laws and behave as ideal gases.

The continuous distillation column is a series of exchanges of heat and mass between liquid and vapor components in a mixture. Each exchange is called a stage which, usually occurs on a tray or on the surface of a packing to insure good contact between high and low boiling components in the mixture. High and low boilers are terms used to describe the components with the higher or lower boiling points as pure substances.

When a mixture of liquids **A** + **B** is vaporized, the vapor becomes more concentrated in the volatile component (low boiler) in direct proportion to its molecular concentration and vapor pressure. This is Raoult's Law at work.

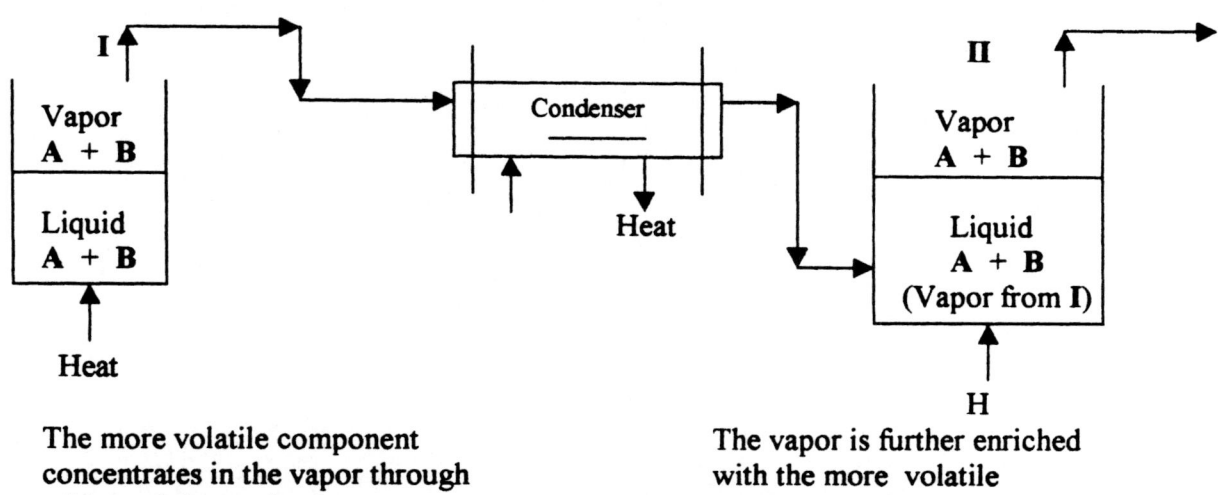

```
    I ↑                                              II ↑
  ┌───────┐                                        ┌───────┐
  │ Vapor │ ──→  ┌───────────┐  ──→               │ Vapor │
  │ A + B │      │ Condenser │                     │ A + B │
  ├───────┤      └─────┬─────┘                     ├───────┤
  │ Liquid│            ↓                           │ Liquid│
  │ A + B │           Heat                         │ A + B │
  └───┬───┘                                        │(Vapor │
      ↑                                            │from I)│
     Heat                                          └───┬───┘
                                                       ↑
                                                       H
```

The more volatile component concentrates in the vapor through with partial vaporization

The vapor is further enriched with the more volatile component when condensate from **I** is vaporized again.

The term volatility is used to describe the ease at which a liquid converts to vapor. Relative volatility is a term used to compare the ease of conversion of one liquid to vapor compared to other liquids in a mixture. The more volatile component is the one with the lowest boiling point.

When the condensate from **I** is partially vaporized again, in **II**, the volatile component is further enriched because of its higher molecular concentration and higher vapor pressure as Raoult's Law predicts.

BASIC DISTILLATION CONFIGURATION

A distillation system contains three basic elements. A reboiler at the base to generate high boiler vapor from a portion of the liquid fed to the column to ensure good vapor liquid contact, and a condenser to condense low boiler vapor.

In addition to the <u>total condenser</u> which condenses both the column overhead make and reflux, a <u>partial</u> condenser is also used. A partial condenser condenses only the liquid used as reflux for the column. In this case the overhead make from the column is a vapor. Note that since the overhead vapor from the column is partially condensed (the liquid and vapor are in equilibrium), the partial condenser serves as an additional tray.

Support equipment such as pre-heaters, pumps, tanks, and column internals are provided to improve the performance of the column as a means to separate miscible liquids.

PROCESS TECHNOLOGY AT WORK IN A
TYPICAL DISTILLATION COLUMN

The mixture of miscible liquids to be separated, called the feed, is introduced in the side of the column. The low boilers overhead are called the make and the high boilers out the bottom are called the tails.

The lower part of the column below the food tray is called the stripping section and the upper part above the feed tray is called the enrichment or rectifying section. The mass and energy exchange between high and low boilers takes place on each tray and a properly designed and operated distillation process will produce the desired purity of low boiler make and high boiler tails.

All materials that enter the column as feed leaves as products. Material balance is an application of the principle of conservation of mass and states:

Input equals output.

Again the principles that make distillation work are:

1. The components in the system are chemically and thermally stable.

2. There is a significant difference between the boiling points and vapor pressures of the components to be separated.

3. The components are miscible. Decantion or settling would be a better choice to separate immiscible components.

4. The desired concentration of low boilers in the make and high boilers in the tails are achieved through use of a practical number of separations (trays or stages) in the column.

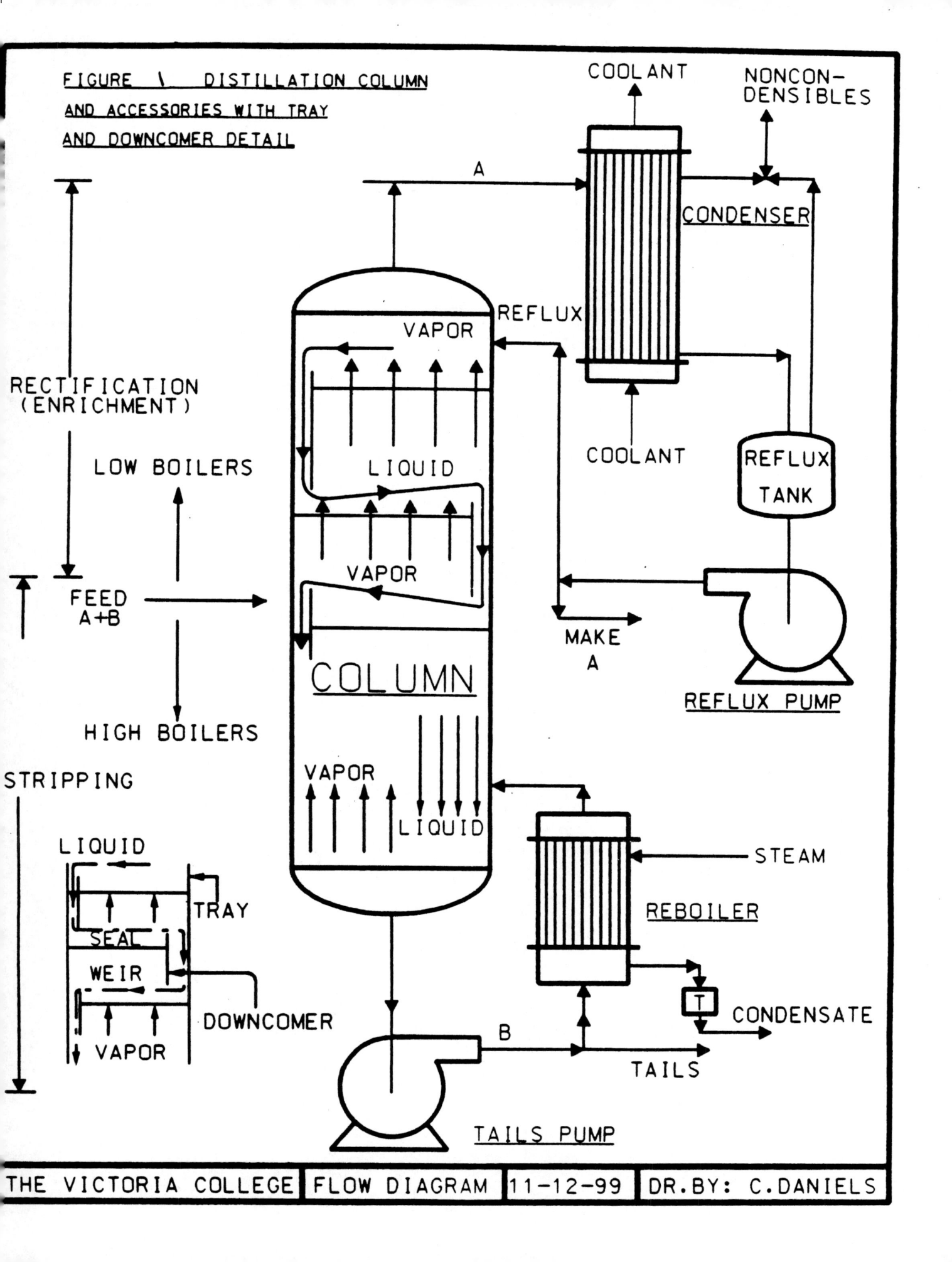

FIGURE 1 DISTILLATION COLUMN AND ACCESSORIES WITH TRAY AND DOWNCOMER DETAIL

| THE VICTORIA COLLEGE | FLOW DIAGRAM | 11-12-99 | DR.BY: C.DANIELS |

FEED RATE

The feed composition should remain relatively constant and should be introduced on the tray with the same liquid composition as the feed. If feed composition changes and is no longer the same as the feed tray liquid composition, the composition of the overhead and bottoms products will change. A column can have several feed points to cover different feed concentrations

The feed rate to the column should also be reasonably constant. A mass balance dictates that the amount of feed into the column equals the sum of the overhead and bottom products. As the feed rate changes the vapor and liquid rates within the column must change. A feed rate too high or too low can lead to inefficient liquid-vapor contacts on the trays, which affect the quality of the separation. The change in flow also has an effect on column temperatures and pressure.

VAPOR RATE

One factor that limits the feed rate to a column is the vapor velocity. Vapor velocity must be high enough to create sufficient pressure drop across each tray to prevent seepage, yet not high enough to excessively entrain liquid to the next tray or to blow the downcomer seal. At high feed rates, the reflux can be reduced to stabilize vapor velocity but at the expense of increased high boilers in the make. If reflux is reduced too far, the separation of high boilers is unsatisfactory.

HEAT INPUT

Move down to the base. Here, in a steam heated reboiler a portion of the liquid is being converted to vapor. The latent heat energy added here creates the vapor flow up the entire column. As the vapor works its way up the column, high and low boilers

contact. The high boilers condense as they provide the latent heat to vaporize the low boilers. The liquid high boilers gravitate down the column and the low boiler vapor moves up the column.

Steam is added to the reboiler to supply the heat necessary to vaporize the liquid. The steam flows around the outside of the tubes and boils the liquid, which is inside the tubes. The vapor from the boiling liquid rises through the vapor line of the calandria into the base of the distillation column. As the steam condenses, the condensate drains to the bottom of the reboiler and is discharged into the sewer through a steam trap. By controlling the heat added to the reboiler, the amount of liquid vaporized can be controlled. Since the vapor from the reboiler passes through the plates to the top of the column the amount of vapor overhead is controlled by regulating the steam added to the reboiler heat.

Energy, in the form of heat, put into the column is also an application of the principle of conservation of energy. That is, the heat put into the column must equal the heat removed from the column. Heat is added to the column by:

1. Steam to the reboiler.
2. Feed stream into column. The temperature of the feed stream affects the product compositions.

Heat is removed from the column by:

1. Cooling water flow from the overhead condenser.
2. Product stream flows.
3. Radiation losses (normally very small and neglected in most cases).

PRESSURE

Pressure has two effects on column operations.

1. An increase in pressure increases the boiling point of a mixture **A** and **B**. therefore the overall operating temperature of a column goes up as pressure is increased. A decrease in pressure lowers the boiling point and therefore the overall operating temperature drops.

2. Pressure increases or decreases vapor density, which has an effect on vapor-liquid contact.

Columns are designed to operate at a fixed pressure in order to maintain constant base and heat temperatures, which are the boiling points of the make and tails. Boiling points set the desired concentrations of the components in the make and tails.

The column pressure is normally automatically controlled by instrumentation to remain constant. If the pressure were not held constant, the quality of both the overhead and bottoms products would change because the boiling temperatures would change.

The column differential pressure is also important. This reflects the amount of restriction to the vapor flow form the bottom of the column to the top. The normal vapor restrictions are:

Flow through each tray opening and flow through the liquid on each tray.
If the differential column pressure is either higher or lower than normal it could indicate possible column interior problems which are discussed later in the Abnormal Operation section.

TEMPERATURE

Feed temperature is best at the temperature on the tray where it is fed to the column. A feed pre-heater often fills this need. Cold feed produces unwanted condensation and hot feed flashes to vapor.

Head temperature is the lowest temperature on the column, and is the temperature of the vapors leaving the column going to the condenser. The head temperature then is essentially the boiling point of the make and defines the composition of the make. Head temperature is used to control the composition of the make, the heat supply to the column, and the reflux to product ration. Obviously, head temperature is a critical control in the distillation process.

Temperature profiles are temperatures measured along the column between the bottom and the top. The temperature profile is recorded to indicate the low and high boiler concentrations. Knowledge of the location of the concentration of the components in the column is a valuable tool in following separation efficiency and column performance.

Side temperatures may be used for the same type control as head temperatures.

The base temperature is measured at or very near the bottom of the column. This is the high temperature point in the column. The base temperature is very important as a measure of tails composition. The base temperature does change with the column pressure.

A high base temperature raises a concern about excessive pressure drop and the effect of high temperature on thermal stability of the tails.

Equilibrium temperature is approached on a middle tray as the cooler liquid from the tray above contacts the warmer vapor from the tray below. If the vapor rate increases, the temperature on the tray increases. Therefore, an increase in heat added to the reboiler increases the vapor rate and increases the temperature on the trays. As the temperature of the tray liquid increases, the liquid will contain more of the heavier component. On the other hand, if the amount of reflux is increased the amount of liquid entering the tray through the downcomer increases. Therefore, increasing the reflux rate lowers the temperature of the liquid on the tray. As the temperature of the tray liquid decreases, the liquid will contain more of the light components.

The temperature profile across a distillation column operating at a fixed pressure represents the boiling points thus the concentrations of components up and down the column. An increase in temperature at constant pressure represents an increase in high boiler concentration and a decrease in temperature at constant pressure represents an increase in low boiler concentration.

REFLUX

The vapor velocity up the column can be stabilized at different feed rates by recycling a potion of the low boilers. This stream is called reflux and serves the primary purpose of improving low boiler purity overhead by knocking high boilers back down the column.

When the product form a distillation contains more of the high boiling material than is desired, an increase in the flow of reflux will usually "wash" this material out of the vapor in the top of the column. The top column temperature is a very good indication of whether the condensate contains any high boiling material.

Reflux is also a means of controlling the temperatures in the column. Increasing the amount of reflux flow reduces the temperatures in the column. Decreasing the reflux flow raises the column temperatures.

Changing the temperature by reflux is simply the result of changing the concentration of high and low boilers. The liquid from the condenser is called condensate. At times a part of this condensate is returned to the top of the column and is called reflux. A column is said to be on <u>total reflux</u> when all the condensate is returned to the top of the column, and when all the condensate is drawn off and none is returned, the column is on total make. Figure 1 shows the system of piping used.

INSTRUMENTATION

Discussion of instrumentation must be general because to specify a type of instrumentation for a distillation column depends largely upon the process. However, instrumentation for a distillation column can be broken down into two categories.

Distillation column control becomes one of maintaining the proper balance of feed rate, reflux rate, and heat supply to give the desired quality in the products. In order to control the variables, easily measurable characteristics must be found to insure proper operation. Temperature control is the most common, but it is not always satisfactory. If the product is of high purity and contains only small amounts of other components, they can vary considerably with no measurable effect on temperature. This is particularly true in the separation of components with close boiling points.

The reflux flow is controlled by the liquid level in the condensate receiver. As the side temperature rises (which indicates an increase in the heavy component) the side temperature instrument positions the overhead make valve closed. This allows more

reflux to be added back into the column, which brings the column temperatures back towards normal. On the other hand, if the side temperature drops (indicating more lights) the make valve is opened. Less reflux is returned to the column and the increased temperatures will drive the lights back up the column.

The particular type of instrumentation chosen for a distillation column depends upon the separation desired. Manual control can be used, but automatic control instruments generally do a better job because, if properly selected, they remove the possibility of over-corrections. In the operation of a distillation column there is normally a long time lag between correction and response. At low pressure, a small change in pressure results in a large change in temperature. For example, from the vapor pressure curve for our component, a column operating at 100mm H_g pressure undergoes a 50mm H_g (or 0.95 psi) change. This pressure change results in a 10°C temperature change in the boiling point. Thus, if the column pressure is not held constant, a change in the column temperature could be mistaken for a change in components.

There are applications where temperatures control the column make. For example, the side temperature instrument positions the overhead make valve. The only manner that a distillation operator can control the purity or quality of the condensed product is to control the amount of reflux returned to the column and in this way control the head temperature.

MORE ON REFLUX

The reflux flow controller attempts to hold the temperature near the top of the column at a fixed value. The steam rate to the still is controlled by a reboiler vapor controller operating on a temperature indicator, and the unvaporized liquid is removed

from the still by a liquid level controller system. The feed flow, reflux flow, and reboiler vapor controllers would be set at the desired values. If the temperature at the top control point became lower than the desired value, the reflux flow controller would reduce the reflux rate, which would increase the product withdrawal rate, which would raise the top temperature. If the temperature were too high, the controller would take the opposite action. If the temperature at the reboiler control point becomes lower than desired, the controller will increase the steam supply, resulting in higher vapor rate. The increased vapor rate will increase the overhead make rate and will probably require an increase in reflux rate to hold the top temperature down. In this system the reboiler vapor controller operates to give the desired temperature at the bottom control point which gives the specified bottom product.

Instrumentation of a distillation column with a detached reboiler or calandria is outlined as follows. A flow rate controller admits the feed to the column at a uniform rate. The feed should be at constant temperature. A pressure controller regulates the release of vapor from the column to maintain a uniform backpressure. A temperature controller controls the steam flow to the reboiler to maintain a uniform temperature of the vapor leaving the reboiler. This method of controlling the steam eliminates the variations due to fluctuations in steam pressure. A flow rate controller on the reflux line regulates the return of condensate to the column. A level controller on the reflux accumulator controls the make flow so that a constant flow of reflux is returned to the column. Instruments record the temperature of the vapor, the feed, the condensate, and the temperature at several locations on the column. Some variations to the above are used. For example, the backpressure on a column can be controlled by regulating the flow of

cooling water to the condenser. Also the flow to the reboiler of a column can be regulated by the pressure differential across the column to maintain a uniform rate of vapor up the column. Ratio controllers can be used to return a fixed percentage of condensate to the column.

Another method of instrumentation of a column is often used. The feed rate to the column is held constant by the feed flow controller. The overhead vapor is condensed and drains into a reflux accumulator. A portion of the condensate is pumped back to the column for reflux. The excess condensate is the overhead make and is removed from the accumulator by the liquid level control system. The reflux rate is controlled to control head temperature and column pressure drop.

Computer applications to distillation process have become common in recent years. A computer is programmed to handle each individual distillation operation. Basically, the computer receives readings from all the column instruments – feed flow, reflux flow, temperatures, etc. Based on these readings the computer calculates what changes are required in the control variables (heat input and flow) to maintain constant quality products. The calculated changes are then compared with specified limits. If the calculated changes are not within in the limits, an alarm is activated to bring the situation to the attention of an operator. (The operator can then check for abnormal operating conditions such as equipment malfunctions, instrument problems, etc.) If the computer-calculated changes are within the limit, the process changes are automatically made. Thus a simple flow diagram (or program) of steps involved in the computer controlled process is shown in Figure 34.

As can be seen the step-by-step process the computer does is basically the same as an operator would do in controlling the column. The advantage of the computer is, however, the speed with which it can analyze the data and make the necessary changes. The decisions of the computer, like the operator, are dependent upon the accuracy of the various instrument readings.

COLUMN INTERNALS

The key to the operation of a distillation column either continuous or batch is the trays or packing which allows contact of the liquid and vapor on each plate. The vapor must be distributed uniformly throughout the liquid on each tray. One method of distributing vapor in the liquid is the use of the bubble cap. To prevent too much liquid from collecting on one plate, a constant level is maintained by adding a weir to the plate. Any excess liquid on the plate overflows the weir and flows through the down pipe to the plate below. The plate also has a weir at the exit of the down pipe so that the liquid from the down pipe will be evenly distributed across the plate.

This will give the bubbles around the bubble caps an opportunity to pass through the liquid being added to the plate. The overflow weir on a plate is placed on the opposite side form the down pipe weir so that the liquid before leaving the plate must pass close to the bubble caps. The down pipe is extended into the liquid on the plate below so the vapor cannot bypass the risers by flowing through the down pipe.

On the bubble cap tray the caps are rigidly fixed to the trays. The Koch flexi-trays have a lift-valve cap which provides a variable tray opening. As the vapor rate through the openings increase, the cap area of the openings increase. As the vapor rate

decrease, the caps start closing. Thus at low vapor flow rates, this type of tray prevents liquid leakage through the openings.

In addition to the bubble cap plates, there are several other plate designs used in industry. Regardless of the type plate used the principle is the same. Provide effective vapor contact with he liquid. Sieve or perforated plate consists of plate with numerous small holes drilled in it to allow the vapor access to the liquid.

ONE COMMON TYPE OF COLUMN IS PACKED

To provide a space where the vapor and the liquid may come in contact for the enrichment to occur, solid materials are frequently used. The liquid flowing down to the bottom of the column flows over the outside of the solids and the vapor rising up the column flows through the spaces or voids between the solids. There is usually sufficient contact between the liquid and the vapor for enrichment of the vapor to occur and therefore, distillation is possible. When a column is filled with solid materials, it is called a packed column.

Special shapes have been designed for column packing and the most common shapes used are called rings. All rings are intended to give good mixing or contact of the vapor and liquid. There are several types of rings, but the rashig ring is most common. The rings may be made of iron, copper, carbon, or ceramic.

The design of a packed column is almost identical with the bubble cap column. Reflux returning to a packed column is added to a reflux distributor which spreads the liquid over the packing and aids in producing good contact between the reflux and the vapor. The distributor is circular and has several V-shaped notches cut out of the sides. As reflux is added to the distributor, liquid overflows the notches or weirs and cascades down through the packing. The packing is supported by a cone shaped screen, which prevents the rings from falling to the bottom of the column. Some packed columns have liquid and/or vapor redistribution plates which helps to overcome "channeling" through the packing. A packed column is cheaper than a bubble cap column of the same capacity but good contact between the liquid and vapor is difficult to achieve when a large diameter column is built. For this reason, a large column is usually a bubble cap column.

AZEOPROPIC MIXTURES

Azeopropes are unique solutions, which are also called constant boiling mixtures. At the specific azeotropic mixture the composition of the liquid is the same as the vapor composition and the boiling point is constant. Thus, azeotropes act as pure materials instead of normal liquid mixtures. Since the vapor composition is identical to that of the boiling liquid, no further purification by distillation is possible.

Normally, azeotropes are either minimum boiling or maximum boiling. Minimum boiling means that the boiling point of the azeotrope is lower than the boiling points of the pure components. Maximum boiling means that the boiling point of the azeotrope is higher than the boiling points of the pure components. In distillation, azeotropes are sometimes used to achieve separations that otherwise would not be possible.

VACUUM OPERATION

Many distillations are carried out under vacuum. This lowers the boiling point of heat sensitive materials and allows vaporization at lower partial pressures of high boilers. Many otherwise impractical distillations can be made successfully under vacuum. The major changes beside temperature are low vapor density and tight limits on allowable pressure drop.

Low vapor density and restrictions on pressure drop result in much larger equipment to process the same quantity of product with adequate holdup time for mass and energy to transfer.

STEAM JET EJECTOR

The steam jet ejector is a common device that is used to pull a vacuum on a distillation operation. The ejector works on the Bernoulli theorem which states that potential energy in a gas created by pressure that is converted to kinetic energy by a reduction in pressure is accompanied by a velocity increase. The high velocity gas will entrain any available material with a loss of pressure in the source of the material.

CRYOGENIC DISTILLATION

Cryogenic distillation at extremely low temperatures is most useful is separating low boiling hydrocarbons such as ethylene and propylene.

Even at very low temperatures, the vapor pressure, thus the operating pressure, is high. Everything from storage to shipping facilities must be kept at very low temperatures to control loss of material. The result leads to very large refrigeration units that operate from -40°F to -80°F that must be highly dependable.

STEAM DISTILLATION

In some operations, the heat input is accomplished by direct steam sparging into the base of the column. This can be done if the water, condensed from the steam, is not harmful to the products or if the steam required will not overload the column, or if added water is actually beneficial to process. These conditions are usually met when the steam provides the heat and vapor to strip residual low boilers from a very high boiling tails stream.

The process is also useful to sparge low boilers from wastewater streams.

BATCH DISTILLATION

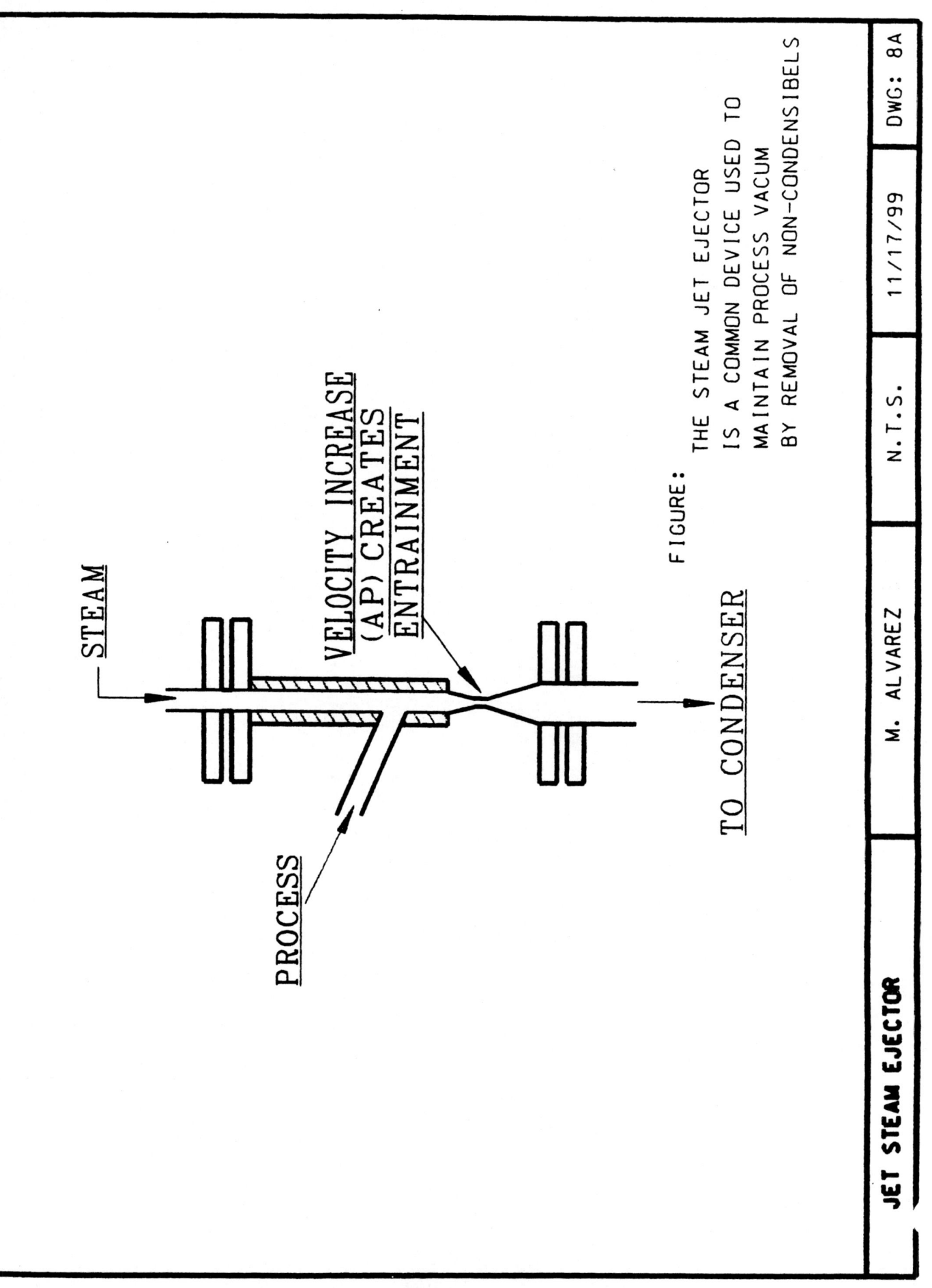

STEAM

VELOCITY INCREASE
(ΔP) CREATES
ENTRAINMENT

PROCESS

TO CONDENSER

FIGURE: THE STEAM JET EJECTOR
IS A COMMON DEVICE USED TO
MAINTAIN PROCESS VACUM
BY REMOVAL OF NON-CONDENSIBELS

| JET STEAM EJECTOR | M. ALVAREZ | N.T.S. | 11/17/99 | DWG: 8A |

At times in a chemical plant the mixtures to be distilled are not available in large or continuous quantities. In these cases only a certain volume of the material is distilled. The quantity handled is usually called a batch and this method of separation by distillation is called <u>batch distillation</u>. Batch distillation requires that the calandria or heating unit be large enough to hold all the batch or charge. This heating unit for batch distillation is usually called a <u>kettle</u>. The kettle is filled with adequate space for the bubbles formed by the boiling liquid.

Batch distillation follows a different procedure than continuous distillation. In batch distillation the material to be distilled is added or charged to the kettle and steam is added through an exchanger to heat the liquid to the boiling temperature. The vapor generated in the boiling liquid serves the same purpose as was explained in continuous distillation. Reflux or liquid collecting overhead can be returned to improve high boiler separation.

In a batch distillation it is possible to separate two, three, or more liquids in a column if they have different boiling points. The mixture is charged to the kettle, the steam turned on to start vaporizing the mixture and the cooling water to the condenser started. Until the material being condensed at the top of the column is certain to be of good quality, the column is kept on total reflux. Since the material **A** boils at a lower temperature than either **B** or **C**, draw-off of make is not started until the top or head temperature is 50°C, indicating that the material is only **A**, which is then recovered. Separation of **B** and **C** follows using the same technique.

Batch type stills generally require little control equipment. They are usually charged with a given quantity of material, which is heated until the volatile components

are removed, and then the still is emptied. The vapor temperature at the top of the still is the main guide. This vapor temperature may be simply recorded while the operator manually controls the heating medium by observation of the temperature recorder. The temperature recorder-controller may control the heating medium automatically or shutoff the heating medium and open the discharge valve when a pre-set temperature is reached.

ABNORMAL OPERATIONS

A tray floods when it is not able to handle either the amount of liquid or vapor. A section of the column is said to be flooded when a group of trays is filled with liquid. A column floods when:

1. The liquid flow in the downcomer is restricted.

2. The vapor flow rate through a tray or a section of trays is too great and "holds up" the liquid. If the liquid cannot flow properly across the tray, there is a buildup in the amount of liquid retained on the tray or trays.

3. The liquid flow rate exceeds the capacity of the downcomers to handle it.

In a column operating normally the composition and temperature of the liquid on each tray is different from that of the liquid on the trays above and below it. If several trays are flooded, the composition and temperature of the liquid is the same on all of them. Thus, all flooded trays act as one normal tray. By reducing the number of effective trays, a flooded section decreases the efficiency of the column and results in inferior quality products.

Under certain adverse operating conditions, trays can become damaged. The bubble caps, flex-caps, etc. can be torn loose and will not provide the proper vapor-liquid contact. The trays may be lifted and tipped out of position by a sudden high vapor

velocity. If they are out of position they will not function properly. A reduction in feed rate should maintain product quality but a column shutdown and repairs made to the trays are the only permanent solution.

Indications of tray damage are same as for dry trays:

1. The temperatures in a section with damaged trays are the same.
2. The column pressure differential will be less if the vapor "bypasses" the damaged trays.
3. Product quality is inferior.

If the column feed contains solids, serious operating problems may develop from tray fouling. Plugging of the tray holes will reduce the effectiveness of the tray.

Indications of fouled trays are:

1. The column pressure differential will be much higher as the vapor flow up the column is restricted.
2. Product quality is inferior.

In sieve tray columns, low vapor rates will allow the liquid to "leak" through the tray holes onto the tray below. Thus, proper vapor-liquid mixing is not achieved. The efficiency of the sieve trays is reduced.

Because of the greater liquid holdup, the pressure drop through a flooded section is much higher than normal. Some indications of flooding are:

1. Column differential pressure is larger than normal.
2. Fluctuation in differential pressure.
3. Small temperature difference in a column section.
4. Off grade products.

To temporarily remedy flooding, reduce both the vapor and liquid flows until product quality returns to that desired.

A distillation column normally operates with a constant level in the base. The bottoms products is withdrawn from the base. To lower the base liquid level, the amount of bottom product is increased. If the level rises too high, liquid will enter the tray section of the column. The rise in the liquid level could be cause by too much liquid entering the column or too little bottoms product being removed. If the liquid rises into the tray section, flooding of these trays will occur. The high liquid level has the same effect as flooded trays previously discussed.

A dry tray is empty of liquid. If there is no liquid on a tray, then no liquid-vapor contact is possible, hence, the tray is not effective. With dray trays the efficiency of the column is decreased. Dry trays can occur if the feed enters the column too hot. The feed, as hot vapor, may contain enough "superheat" to vaporize the liquid on the trays just above the feed tray. Dry trays may also occur in the stripping, or lower section and are caused by superheated vapor.

Indications of dry trays are:

1. The temperatures on a section of dry trays are the same.
2. The column pressure differential will be less as the vapor has less liquid to overcome.
3. Loss of column efficiency and inferior product quality.

To correct dry trays the feed temperature can be reduced or the reflux to the dry section of the column can be increased.

GENERAL SAFETY CONSIDERATIONS

Attack of the metal and an extra thickness of metal is used to allow for some corrosion. Corrosion should be carefully watched as it could cause the column shell to fail.

The column shell is designed to withstand the normal operating pressure plus some safety factor. Normally the design pressure is 1.5 times the normal operating pressure. To protect the column, pressure relief devices and rupture discs are installed. These relief devices are set at some pressure between the normal operating pressure and the design pressure. It is unsafe to valve-in the relief devices or allow them to become clogged during column operation.

Metals lose a potion of their strength when heat is applied. Metals expand under heat. The expansion must be considered in the column design or it can cause damage. All large equipment should be heated and cooled at a rate to control expansion and contraction.

If the material involved in a distillation process is explosive, care must be taken to keep the material from contacting air. For processes in which explosive materials are used, inerts (such as nitrogen) are continuously added to the column vent or vapor system to reduce the possibility of explosive mixtures. Unless special care is taken, explosive mixtures are more likely during column start-up and shutdown.

EQUILIBRIUM refers to a condition where mass or energy into and out of a fixed point is the same.

BATCH is a freestanding amount of material.

AZEOTROPE is a mixture that has the properties of a single component, i.e., constant vapor pressure, constant boiling point, and etc.

TERMS AND SYMBOLS

<u>Distillation</u> is a process that utilizes differences in boiling point and vapor pressure to separate miscible components in a mixture by repeated partial vaporization and condensation with separate recovery of the liquid and vapor.

<u>Steady State</u> is an operating condition where all variables stay within normal values over time.

<u>Stripping</u> is the name commonly given to the section of a distillation column below the feed point.

<u>Rectification</u> is a common form of distillation where the vapor is in continuous contact with the liquid that permits exchange of heat and mass. It is also the name commonly given to the section of a distillation column above the feed point.

<u>Reflux</u> is the return of a portion of the condensed vapor back to the column to improve rectification. It can also be used to maintain column stability (boil-up).

<u>Stage</u> is one contact of liquid and vapor usually on a tray or the surface of a packing such as Beryl saddles.

<u>Vapor Pressure</u> is the pressure exerted by the vapor in contact with the host liquid. It increases with temperature and molecular concentration of the component in the liquid.

<u>Partial Pressure</u> is the contribution to the total pressure by each component in the host liquid.

<u>Miscible</u> is a term used to describe a mixture of liquids that form a single phase in all proportions.

<u>Boiling Point</u> is the temperature that liquid converts to vapor called a phase change.

<u>Density</u> is the weight per unit volume of a substance.

Cryogenic refers to operations far below normal atmospheric conditions.

Immiscible is a term used to describe a mixture of liquids that form two phases i.e. do not mix in any proportion.

Topper is a column that primarily removes highly volatile components.

Refiner is a column that produces the finished product.

Concentration Still is a column used to concentrate a component, usually by removing water.

Demethanizer, De-ethanizer, Depropranizer, etc. are columns that remove a specific chemical from a mixture.

High Boilers are the less volatile components in a mixture and come out in the tails.

Low Boilers are the more volatile components and go overhead.

Downcomer is an enclosed space that conveys liquid from tray to tray.

Steady State is the condition where mass and energy transfer into and out of a system are at equilibrium.

Weight Concentration, Wt. % or Wt. frac. Is simply the weight of a component divided by the total weight of a mixture. Weight concentrations are used in solutions, mixes and etc. One pound of **A** mixed with one pound of **B** is a 50 wt. % mix of **A** or **B**.

Molecular Concentration mol. % or mol. frac. Is the moles of a component divided by the total moles in a mixture.

Reboiler is a heat exchanger that supplies heat energy to the distillation column by evaporation of a circulating tails stream.

Tails is the material removed from the bottom of the column.

<u>Feed</u> is the mixture to be separated and is fed to the column at the point of matching concentrations.

<u>Make or Heads</u> is the potion of the overhead vapor that is condensed and removed. A portion is recycled as reflux.

<u>Condensate</u> is overhead vapor converted to liquid.

<u>Products</u> are the components or component recovered for additional use. It can be the tails, make, or a side stream.

<u>Volatility</u> is the comparison of the vapor pressures of two or more substances at constant temperature.

CHAPTER 8

WATER TREATMENT

One of the major attractions for locating petrochemical plants on the Gulf Coast is a plentiful supply of surface water. Water can typically be taken from a river into the plant at a point upstream, used and then discharged back into the river downstream. Use of natural water requires adequate environmental precautions not to overheat or contaminate the source.

Two sources of water are available for industrial plant use. These waters are very different in the impurities they contain. There is surface water such as rivers and lakes. As a rule the user can expect:

- Low in dissolved solids
- High in suspended solids
- Quality changes caused by the seasons and weather

The second source of industrial water comes from wells.

As a rule, well water is:

- High in dissolved solids
- Low in suspended solids
- High iron and manganese content
- Low oxygen but may contain sulfide gas
- Relatively constant quality and temperature

The dissolved solids in both can cause scale and corrosion.

Waters is a chemical with unique properties (See Table 1). Its use in the Petrochemical Industry to cool, heat, prepare solutions, rinse, and quench is a versatility

no other chemical can match. Natural water is filtered, chemically softened, and de-mineralized in stages, as necessary, to meet plant purity requirements. Pure water is non-corrosive, non-toxic, and non-flammable. This makes water uniquely safe for use in industrial processes with a low hazard potential to personnel Water is very stable chemically, so does not take part in most chemical reaction mechanisms. It is the solvent of choice in the production of many process solutions. Acids and bases are examples.

Natural waters are not pure water (See Table 2). Natural water is a solution of a number of chemicals and contains suspended solids. Some of these dissolved chemicals are corrosive to metals and form scale deposits on metal surfaces. This is unacceptable for continuous plant use in cooling applications. Also, water for use in contact with process materials usually requires a high level of purity.

The removal or reduction of the impurities in natural water is called <u>Treatment</u> (See Block Diagram). Water treatment does not produce chemically pure water, which is rarely attainable by any means. The objective of water treatment is to reduce the level of impurities to an economical concentration that is acceptable for the end use.

Raw water impurities vary widely with the source, flow rate, and the terrain from which it drains. Metal, salts, and suspended solids are the most harmful impurities. Odor and taste are important considerations for potable use but is not particularly important in industrial applications.

Well water is still available for industrial use along the Gulf Coast but demand is fast approaching the available supply. Since municipalities often take priority over industrial and agricultural users, reducing water consumption is an important plant

priority. Well water is usually clear, odorless, and suitable for potable use without treatment. The primary objective is to remove dissolved salts that produce hardness. Once rainwater was allowed to drain naturally from industrial processes, or it was directed by man-made drainage systems into holding ponds and out-falls. Today, most plants collect rainwater from process areas and check for contamination before it is discharged to the river. If necessary, the water is treated to insure acceptable purity levels before discharge to prevent possible contamination of a natural source.

TABLE 1

PROPERTIES OF WATER

Weight, Volume, and Density Constants

62.4 pounds per cubic ft.

1 gram per milliliter

1.0 specific gravity

8.3 pounds per gallon

7.48 gallons per cubic ft.

212°F at 14.7 psia	boiling point
32°F	freezing point
*1,000 BTU per pound	latent heat
1.0 BTU per pound per °F	heat capacity
1.0 calories per gram per °C	heat capacity
1.0 at 25°C	specific heat

*### Vapor Pressures at 212°F (100°C)

760mm H_g, 14.7 psia, 29.92 in. of H_g and 33.4 ft. of water at the boiling point of 212°F, 100°C. Varies with boiling point and is equal to the pressure of the system in which vaporization is occurring.

TABLE 2

TYPICAL IMPURITIES IN NATURAL WATER

Component	Chemical Base	Problem Cause	Treatment
Hardness	Calcium and magnesium salts reported as Calcium Carbonate, $CaCO_3$ equivalent	Scale in heat exchangers	Chemical removal usually by soda ash process
Alkalinity	Bicarb (HCO_3) OH ion reported as $CaCO_3$ equivalent	Foaming, solids in steam, generation embrittlement. Forms CO_2 in steam which is corrosion	Lime and lime-soda softening. Demineralization. Distillation for very high purity.
Mineral Acids	Sulfuric (H_2SO_4), Hydrochloric (HCl) and etc. reported as $CaCO_3$ equivalent	Corrosion	Neutralization
Carbon dioxide Oxygen	CO_2 O_2	Corrosion, particularly in steam systems	Aeration and deaeration
Color Turbidity	None None	Suspended solids produce deposits on metal surfaces	Coagulation and setting (usually removed with hardness)

The major impurities of concern in natural water sources cause either scale or corrosion. Some of them such as oxygen, carbon dioxide, and hydrogen sulfide are degradation products of organic matter.

The permanent hardness can be attributed to the metallic salts of magnesium and calcium. Softening is a process to remove the metal ions and replace them with a "soft" ion such as sodium (Na+).

Alkalinity is caused by the impurities bicarbonate (HCO_3) and hydroxyl ion (OH). They are precipitated by lime addition and purged to prevent metal embrittlement.

FILTRATION

Crude filtration is usually the first step in water treatment. Its purpose is to remove large suspended solids. Small particles are removed by coagulation prior to the softening process. There are screens on river intakes to keep large floating objects out of pumps, piping, and canals. The simplest form of solids removal is a settling basin or pond where dead time is provided for large particles to settle out. There is little effect on small-suspended particles and no effect on colloidal particles. A settling basin also has merit in providing holdup time to level out large swings in suspended solids caused by rain run-off and river flood conditions.

A sand or gravel filter is a common way to remove intermediate sized particles. An activated carbon bed will do an effective job on removing small particles and has the ability to improve odor and taste. However, it is impractical for use on very large quantities.

REMOVAL OF SUSPENDED SOLIDS

Clarification is the removal of suspended matter and color from water supplies. The suspended matter may consist of large particles that settle out readily in these cases, clarification equipment merely involves the use of settling basins or filters. Most often, suspended matter in water consists of particles so small that they do not settle out, but instead pass through filters. The negative charge on the small particles causes them to repel one another. The removal of finely divided or colloidal substances requires the use of a coagulant.

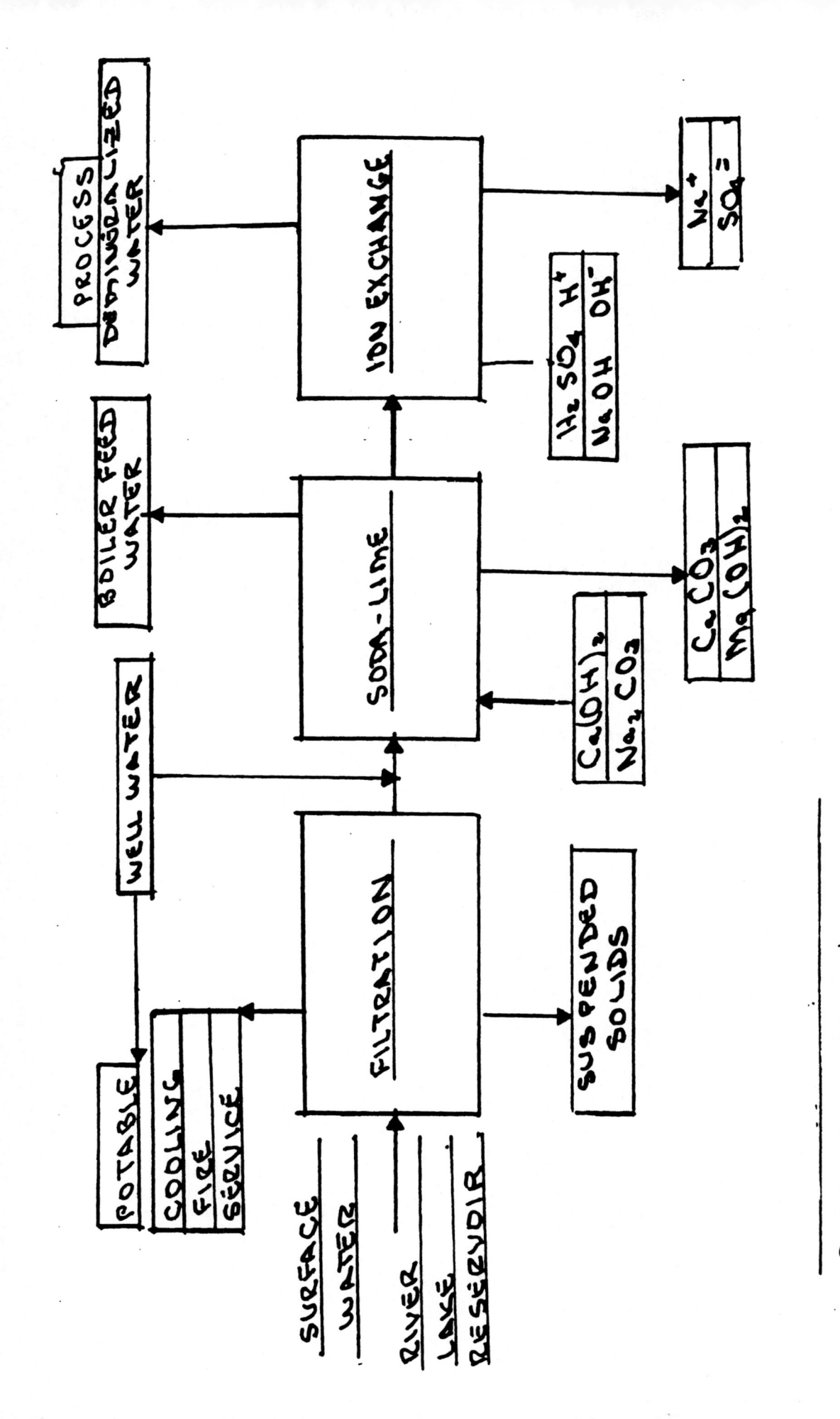

Block Diagram 1
Water Treatment

Surface Water (River, Lake, Reservoir) → **FILTRATION** → Suspended Solids

Filtration outputs: Potable, Cooling, Fire Service

Well Water → **SODA-LIME** (with Ca(OH)$_2$ / Na$_2$CO$_3$ added) → CaCO$_3$ / Mg(OH)$_2$

Soda-Lime output: Boiler Feed Water

→ **ION EXCHANGE** (with H$_2$SO$_4$, H$^+$, NaOH, OH$^-$) → Na$^+$ / SO$_4^=$

Ion Exchange output: Process Demineralized Water

T.B.F. 7/60

COAGULATION AND FLOCCULATION

Coagulation involves neutralizing the negative charges and provided a nucleus for the suspended particles to adhere to. Flocculation is the bridging together of the coagulated particles.

The most common coagulants are iron and aluminum salts such as ferric sulfate, ferric chloride, aluminum sulfate (alum), and sodium aluminate. Ferric and alumina ions each have three positive charges and therefore their effectiveness is related to their ability to react with the negatively charged colloidal particles. With proper use these coagulants form a floc in the water that serves as a kind of net for collecting suspended matter. Synthetic materials, called polyelectrolytes, have been developed for coagulation purposes.

SOLID LIME PROCESS

The single most important family of impurities in natural water that require treatment is the salts and hydrates of magnesium and calcium metals. These two metals are a major source of fouling and scale formations on equipment surfaces and their hydrates are corrosive. The two processes frequently used in industry to soften water are soda-lime and ion exchange.

The soda-lime process is a two chemical attack that converts the soluble salts and hydrates to insoluble carbonates, which precipitate and settle out for removal as a purge stream.

In precipitation processes, the chemicals added react with dissolved minerals in the water to produce a relatively insoluble reaction product. Precipitation methods are used in reducing dissolved hardness, alkalinity, and in some cases silica. The most common example of chemical precipitation in water treatment is lime-soda softening.

LIME TREATMENT TO REMOVE TEMPORARY HARDNESS

Hydrated lime reacts with calcium and magnesium bicarbonates to form insoluble salts that precipitate. The bicarbonate ion is identified with temporary hardness.

$Ca(OH)_2$ +	$Ca(HCO_3)_2$	$2\,CaCO_3$ +	$2H_2O$
Hydrate of lime	Calcium Bicarbonate	Calcium Bicarbonate	Water

$2\,Ca(OH)_2$ +	$Mg(HCO_3)_2$	$Mg(OH)_2$ + $2\,CaCO_3$ +	$2H_2O$
Hydrate of lime	Magnesium Bicarbonate	Magnesium Hydroxide / Calcium Carbonate	Water

The precipitants calcium carbonate ($CaCO_3$) and magnesium hydroxide $Mg(OH)_2$ are sludge like materials that can be removed by settling and blowdown in a clarifier (See Figure). Often, coagulation is required since the participant can be quite small.

SODA LIME PROCESS

The single most important family of impurities in natural water that require treatment is the salts and hydrates of magnesium and calcium metals. These two metals are a major source of fouling and scale formation on equipment surfaces and their hydrates are corrosive. The two processes frequently used in industry to soften water are soda lime and ion exchange.

The soda-lime process is a two chemical attack that converts the soluble salts and hydrates to insoluble carbonates, which precipitate and settle out for removal as a purge stream called blowdown.

In precipitation processes the chemicals added react with dissolved minerals in the water to produce a relatively insoluble reaction product. Precipitation methods are used in reducing dissolved hardness, alkalinity, and in some cases silica. The most common example of chemical precipitation in water treatment is lime-soda softening.

LIME TREATMENT TO REMOVE TEMPORARY HARDNESS

Hydrated lime reacts with calcium and magnesium bicarbonates to form insoluble salts that precipitate. The bicarbonate ion is identified with temporary hardness.

$$Ca(OH)_2 + Ca(HCO_3)_2 \rightarrow \downarrow 2 CaCO_3 + 2H_2O$$

Hydrate of Lime + Calcium Bicarbonate → Calcium Bicarbonate + Water

$$2 Ca(OH)_2 + Mg(HCO_3)_2 \rightarrow \downarrow Mg(OH)_2 + 2 CaCO_3 + 2H_2O$$

Hydrate of Lime + Magnesium Bicarbonate → Magnesium Hydroxide + Calcium Carbonate + Water

The precipitants calcium carbonate ($CaCO_3$) and magnesium hydroxide $Mg(OH)_2$ are sludge like materials that can be removed by settling and blowdown in a clarifier (See Figure). Often, coagulation is required since the precipitant can be quite small.

SODA ASH TREATMENT TO REMOVE PERMANENT HARDNESS

Soda ash reacts with calcium and magnesium salts to form carbonates. Calcium sulfate converted to calcium carbonate is used to demonstrate the chemistry.

Na_2CO_3 + $CaSO_4$ → $CaCO_3 \downarrow$ + Na_2SO_4
Soda Ash   Calcium Sulfate   Calcium Carbonate   Sodium Sulfate

Na_2CO_3 + $CaCl_2$ → $CaCO_3 \downarrow$ + $2NaCl$
Soda Ash   Calcium Chloride   Calcium Carbonate   Calcium Chloride

The precipitant, calcium carbonate, (and other solids), are removed using coagulation and flocculation to aid settling followed by blowdown/purge. Any residual hardness in the water after softening is reported as ppm $CaCO_3$ (Calcium Carbonate) regardless of the true chemical nature of the hardness. The actual quantity of hardness other that $CaCO_3$ is calculated by the ratio of the molecular weights multiplied by the reported hardness as $CaCO_3$.

The soda-lime process is effective in removing both temporary and permanent hardness but is not trouble free. A few things to keep in mind.

1. The chemical reactions are slow and carried out in very low concentrations. A typical example would be:

Na_2CO_3	+	$CaSO_4$	→	$CaCO_3 \downarrow$	+	Na_2SO_4
106		136	molecular wts.	100		142

Heats of Formation kcal/g. mol

-269.5   -338.7   -290.0   -330.5

-269.5   -290.0
-338.7 Reactants   -330.5 (Products)
-608.2   -620.5

The sum of the heats of formation of the products minus the sum of the heats of formation of the reactants equals the heat of reaction.

323

620.5 - 608.2 = 12.3 k cal / g. mol
12.3 x 1.8 = 22.1 M BTU/lb. mol. of Ca CO$_3$ produced.

The molecular weight of soda ash (Na$_2$ CO$_3$) is 106.

$$\frac{22.1 \text{ M BTU/lb. mol}}{106 \text{ lb./lb. mol}} = 208 \text{ BTU/lb. of soda ash reacted.}$$

Suppose there is 500 ppm of hardness measured as calcium carbonate (CaCO$_3$) in the feed water to be removed. Under ideal conditions, it would take:

$$\frac{106 \text{ lb./ lb. mol Na}_2\text{CO}_3}{100 \text{ lb./lb. mol Ca CO3}} \times 500 = 530 \text{ ppm of Na}_2\text{CO}_3 \text{ to remove 500ppm of CaCO}_3.$$

This can also be stated as 530 lbs. of Na2 CO3 is stoichemetrically required to remove 500 lbs. of CaCO3 from 1,000,000 lbs. of water.

530 lbs. of Na2CO3 x 158 BTU/lb. = 837,400 BTUs per million lbs. of water.

The temperature rise from the reaction would be:

$$\frac{837,400 \text{ BTU}}{(1,000,000 \text{ lbs.})(1 \text{ BTU/lb./°F})} = 0.83°F \ \Delta T$$

The 0.83°F temperature rise is considered negligible.

Water softening systems are often heated to speed up the chemical reactions and excess chemicals are used to drive the reaction to completeness.

COAGULATION –FLOCCULATION TO REMOVE PRECIPITANTS

Coagulation is often needed because the suspended solids formed in soda-lime treatment usually carry a negative charge and repel each other. The precipitate would settle slowly if ever without the formation of larger conglomerates. Coagulants can speed up settling of sludge as much as 25 –50%.

Coagulation involves the introduction of ferric sulfate, which is positively charged, to neutralize the negatively charged particles. The neutralized particles will collide among themselves in the reaction zone of the clarifier to form a larger particle known as floc. Because pH can affect both particle surface charges and floc precipitation during coagulation, it is important that the pH be controlled from 9.8 to 10.1. There is no control test for the ferric sulfate concentration of the clarifier. This is a typical example where operator experience and judgement are important. Underfeed of ferric sulfate is noticed by poor floc formation in the reaction zone and a milky look to the clarifier effluent. Overfeed of ferric sulfate will also produce poor floc formation in the reaction zone, but the clarifier effluent will have a reddish tint to it.

REFINED FILTRATION

Filters are used to remove any turbidity left in the water after clarification. The flow rate through the filters is controlled and totaled in order to monitor throughout. When a preset amount of water has passed through the filter, it is switched to a clean unit. A common practice is to switch and backwash filters on a timed, routine cycle. The filters are cleansed of the filtered matter by backwashing with a flow opposite to the filtration flow. The filter is then rinsed for a preset time and is ready for service. In short, at least two filters are needed to operate a continuous process.

Often water is treated with chlorine to destroy bacteria. Sodium sulfite is added to the filter effluent to react with the residual chlorine. The target for chlorine residual is 0.0 ppm. Chlorine will attack the cation resin in ion exchange and decompose it. The decomposed cation resin will then foul the anion resin. Excess sodium sulfite will add some load to the cation and anion beds, but it would take a large overdose to significantly affect the bed capacity.

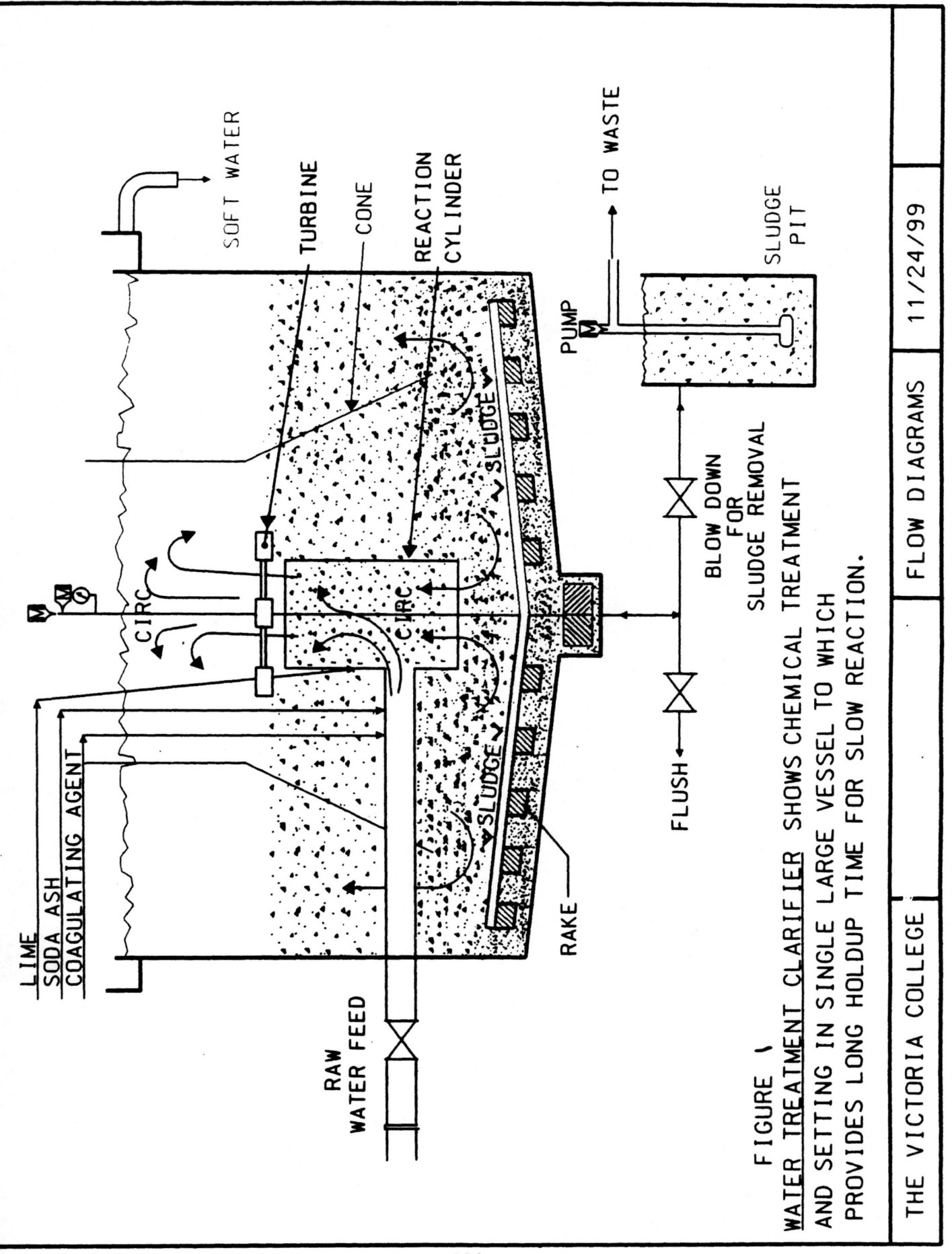

FIGURE 1
WATER TREATMENT CLARIFIER SHOWS CHEMICAL TREATMENT AND SETTING IN SINGLE LARGE VESSEL TO WHICH PROVIDES LONG HOLDUP TIME FOR SLOW REACTION.

THE VICTORIA COLLEGE | FLOW DIAGRAMS | 11/24/99

ADSORPTION

Adsorption is the process in which an impurity is removed from a process stream by making it adhere to the surfaces of a solid.

Adsorption processes exist in a variety of forms, which include ion exchange, molecular sieves, silica gel, and activated carbon. These processes are used in widely varying applications including softening.

The process depends upon the electrical charge on ions. Ions are atoms that have gained or lost electrons in their outer orbits and have either a positive or negative charge. A positively charged ion is called a "cation". Negatively charged ions are called "anions". Ion exchange resins on which the absorption takes place are classified as "cationic" if they remove cations, and "anionic" if they remove anions. Ion exchange resins are very small, bead-like particles.

In an ion exchange process, the process stream containing the impurities, which are in an ionized form, is passed through a bed of the ion exchange resins. Water treatment is a good example. "Hard" water contains salts of metals such as calcium, magnesium, or iron. Water-treating resins have sodium ions active on their surfaces. The sodium ion replaces the "Hard" ion, the latter remains attached to the surface of the resin bead.

As in the case of the ion exchange resins, a point is reached when all the pores are filled and the bed is saturated. At this point it must be regenerated. This can be done by a wash with a solution of the small ion.

Equipment for adsorption operations is relatively simple. It consists primarily of a tank or vessel containing a "bed" of the adsorbent, be it ion exchange resins, molecular

sieves, or whatever. In most cases the bed is "fixed." That is, the bed is held in place while the process stream containing the impurities is passed through the bed. In a few cases, the beds are "fluidized" and are passed through the vessel counter-currently to the flow of the process stream. In such instances, the adsorbent particles are separated as they leave the vessel and are passed through a regeneration step. The regenerated particles are then recycled to the system.

The fixed bed type are by far the more common adsorbers. They contain some type of bed support mechanism usually a screen pack to hold the bed in the vessel. This would typically consist of sort of grid support, covered by wire-mesh screens, In some cases the screen is then covered with graduated sizes of gravel or other solid particles to support the adsorbent. Such support facilities are needed to prevent the very small adsorbent particles from being washed out of the bed and to distribute the flow evenly across the bed. The process stream is usually fed through some sort of pipe sparger or distributor to further assure even distribution across the cross-sectional area of the bed. The process stream is removed through a similar type device, usually covered with a wire-mesh screen to avoid loss of the particles.

SOFTENING WATER BY ION EXCHANGE

Ion exchange softening is commonly dependent on the sodium (Na^+) cycle (See Figure). Resins, called zeolytes, have the physical property to attract and hold positive ions. In softening water, the resin is loaded by Na^+ ions by regenerations with a salt (sodium chloride) solution.

Water containing metal ions, such as Mg^{++} and Ca^{++}, are passed through the bed and displace the Na^+ ions on the surface of the zeolyte. When the bed is loaded with Mg^{++} and Ca^{++} ions, the flow is diverted to a freshly regenerated bed. The loaded bed is regenerated with a sodium chloride solution. The high concentration of Na^+ ion in the solution displaces the Mg^{++} and Ca^{++} ions, which are purged. The bed is regenerated and ready for service. The length of the cycle depends upon the size of the bed, the concentration of impurities, and the total feed.

Channeling is a potential problem with flow beds of through solids. Flow naturally takes the path of least resistance or lowest pressure drop per unit of flow. This means more flow will pass through an area of less resistance at the same pressure drop.

Channeling creates several process problems.

1. The area of high flow is loaded preferentially and leads to early breakthrough at a less than normal loading.

2. The area of high flow provides for less time for ion exchange to take place that can lead to early breakthrough and low total loading.

The key to good distribution of flow is proper installation of the resin bed and additional pressure dorp. Screens are provided at the top and bottom of the bed to hold

the resin in place and to create significant pressure drop compared to the drop across the bed. This makes the pressure drop across the bed less a part of the total.

Depending upon the use for which it is intended, an ion exchange resin may have any one of a number of different ions active on its surface. Hydrogen ions are a common type that find application in chemical processes. Sodium ions are convenient in small softening units where sodium in the product is acceptable.

FIGURE

THE SODIUM CYCLE

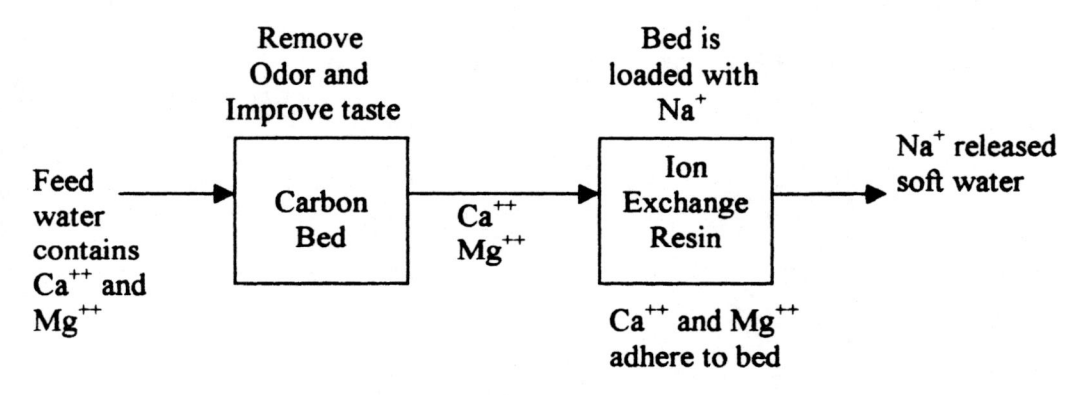

```
Feed water              Remove                Bed is
contains                Odor and              loaded with
Ca++ and   ──▶  Carbon Bed  ──Ca++──▶  Ion Exchange  ──▶  Na+ released
Mg++                    Improve taste  Mg++    Resin           soft water

                                              Ca++ and Mg++
                                              adhere to bed
```

ON LINE LOADING

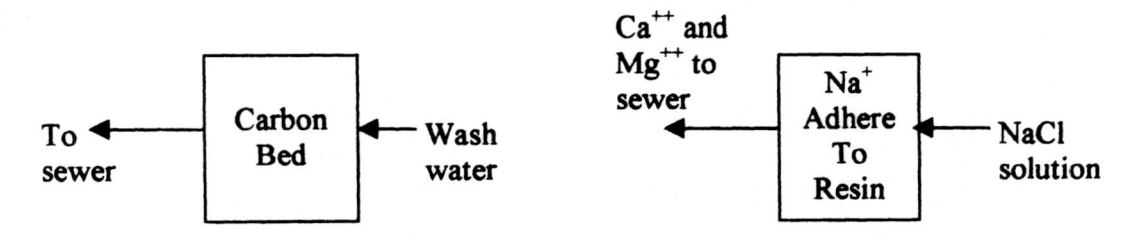

```
                                              Ca++ and
                                              Mg++ to
                                              sewer
To sewer ◀──  Carbon Bed  ◀── Wash water ──  Na+ Adhere To Resin  ◀── NaCl solution
```

REGENERATION

DEMINERALIZATION

After the hardness has been removed, there are still foreign chemicals in the water. Though the concentration of impurities that cause hardness is low after treatment, some processes require water with a low concentration (high purity) of all contaminants that effect product quality. Often high purity water is needed to prepare solutions or dissolve process gases to produce acids.

The most common ions left in the water are the cation Na^+ and anions such as So_4^- from the soda ash. The process to remove these ions is called demineralization, which is accomplished through use of ion exchange resins.

For continuous operation, there are two complete ion exchange beds or units needed to remove the undesirable cations$^+$ and two complete ion exchange units needed to remove the undesirable anions$^-$. One unit is being regenerated or on standby while the other is in process service.

Demineralization is accomplished by a process called ion exchange, which is the adsorption, and de-adsorption of selected materials on an activated solid surface. The objective is to trade undesirable cations for a useful cation and trade undesirable anions for a useful anion. The solid surface ion exchange resins are usually in pellet form and installed in fixed beds.

Ion exchange resins hold either a high positive or a high negative charge and attract ions with an opposite charge.

High Positive   **High Negative**

The high positive resin attracts and holds negative ions and the high negative resin attracts positive ions. The process is called adsorption. When the ion on the resin is replaced by an ion that is lower on the elctromotive series, an exchange has occurred.

CATION REMOVAL PROCESS (Figure 3)

The objective is to remove cations such as Na^+, Fe^{++}, and replace them with H^+. This is accomplished by passing the water through an ion exchange bed saturated with H^+. the Na^+ and other cations are adsorbed and displace an H^+ from the negatively charged resin. The H^+ goes out with the water.

The H^+ ions will continue to be displaced by the Na^+ ions until the bed is spent; i.e. The bed contains all the Na^+ ions it can hold and Na^+ appears in the treated water. This is called breakthrough and the bed is said to be loaded. In practice, through put is monitored and the beds are switched before breakthrough occurs (See Figure 5).

The flow is switched to a second bed that has been washed with a strong acid solution to remove all the Na+ ions and replace them with H+ ions (See Figure 2A). This process is called regeneration.

The alternating on line and regeneration steps provide a continuous supply of harmful cation free water.

Anion Removal Process (Figure)

The operation of the two anion removal beds is exactly the same as the cation beds. The difference is only in the adsorption where we exchange undesirable anions like SO4- (sulfate) with hydroxyl (OH-) ions. The net effect is:

$$Na^+ \text{ replaced by } H^+$$
$$SO_4^- \text{ replaced by } OH^-$$
$$H^+ + OH^- \rightarrow H_2O$$

ON LINE EXCHANGE ANION REMOVAL

SUPPORT SCREENS

FEED WATER
Na SO$_4$ →

→ WATER PRODUCT

ADSORPTION
1. $SO_4^=$ RETAINED
2. OH^- RELEASED

RESIN BED

REGENERATION ANION REMOVAL

SUPPORT SCREENS

REGENERATION FLUID
Na OH →   WASH

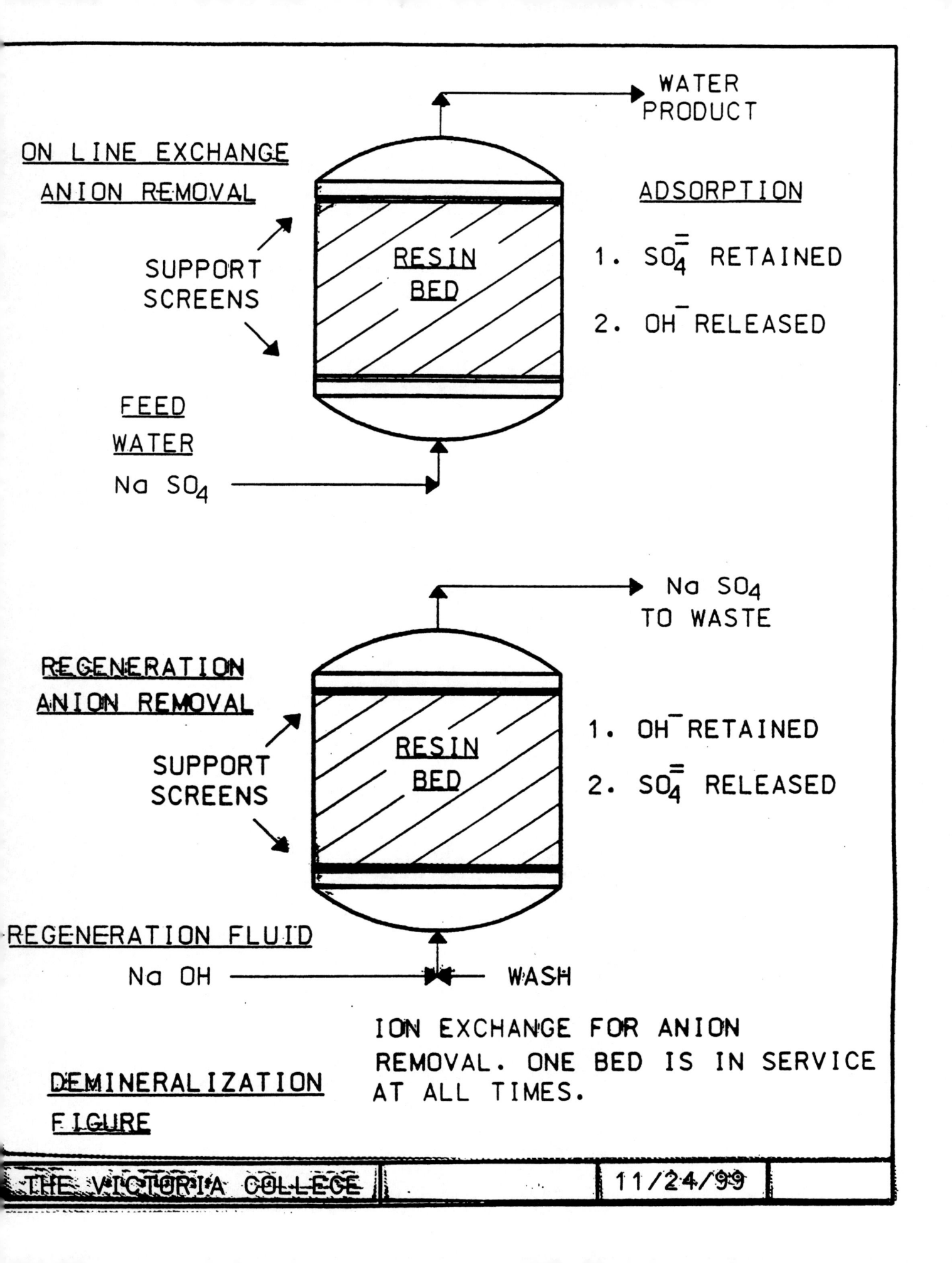

→ Na SO$_4$ TO WASTE

1. OH^- RETAINED
2. $SO_4^=$ RELEASED

RESIN BED

ION EXCHANGE FOR ANION REMOVAL. ONE BED IS IN SERVICE AT ALL TIMES.

DEMINERALIZATION FIGURE

THE VICTORIA COLLEGE — 11/24/99

A.

→ PRODUCT
WATER WITH H_2SO_4
FED TO ANION REMOVAL

SUPPORT SCREENS → RESIN BED (H^+)

1. Na^+ RETAINED BY ADSORPTION
2. H^+ REPLACED Na^+

FEED WATER
Na_2SO_4
TO BE REMOVED

B.

→ Na_2SO_4 TO WASTE

REGENERATION CATION REMOVAL

SUPPORT SCREENS → RESIN BED

1. H^+ RETAINED BY ADSORPTION
2. Na^+ REPLACED

REGENERATION FLUID
H_2SO_4 5%+ SOLUTION — WASH

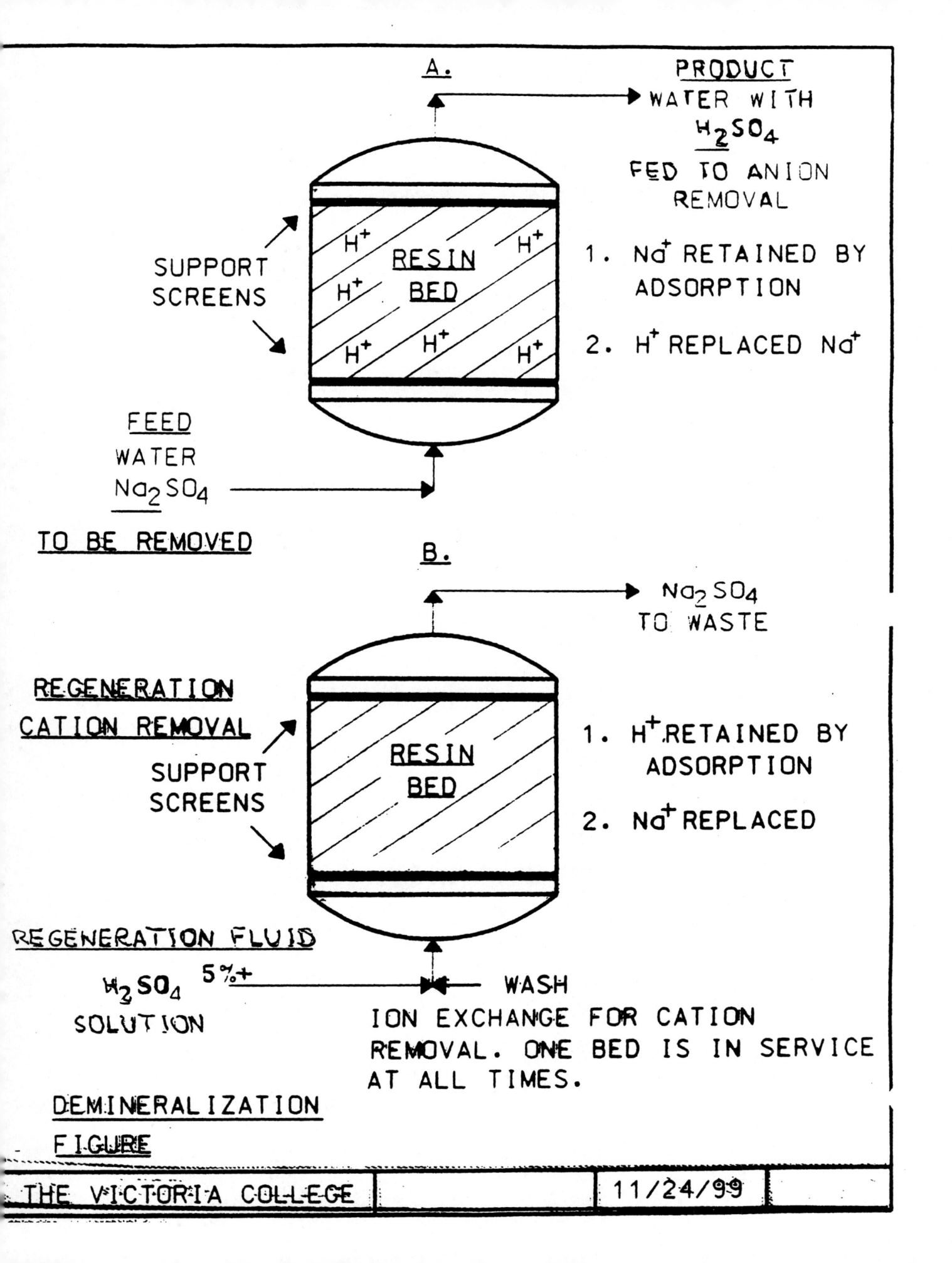

ION EXCHANGE FOR CATION REMOVAL. ONE BED IS IN SERVICE AT ALL TIMES.

DEMINERALIZATION FIGURE

THE VICTORIA COLLEGE — 11/24/99

PROBLEMS WITH DEMINERALIZATION

Resins do lose capacity with throughput and time. The most obvious resort is a need for more frequent regenerations. Eventually, the bed must be replaced to maintain plant capacity.

Channeling creates several process problems:

1. The area of high flow is loaded unevenly and leads to early breakthrough.

2. The area of high flow provides for less time for ion exchange to take place and also can lead to early breakthrough.

The keys to good distribution of flow is proper installation of the resin and pressure drop. Screens are provided at the top and bottom of the bed to hold the resin in place and to create significant pressure drop. This makes the pressure drop across the bed less of the total.

Low feed rates result in low pressure drop and poor mixing. Channeling is more prevalent at less than normal feed rates.

TERMS AND SYMBOLS

Latent Heat – The amount of heat needed to convert a pound of a substance to a higher energy phase. Heat to convert ice to water and water to steam are two examples. The process is reversible.

Boiling Point – The temperature at which a liquid converts to vapor at a constant pressure. Boiling point temperature increases with increased system pressure and decreases as system pressure decreases.

Heat Capacity BTU/lb./°F – The heat required to heat one pound of a material (liquid, gas, or solid) one degree °F.

Hardness – A term used to describe water in which the cleaning power or suds formation of soap is poor. Poor performance is caused by reaction of the soap with the chemicals (impurities) in the water. As an international standard, hardness is measured as equivalent calcium carbonate (Ca CO3) concentration in parts per million, ppm.

Soft Water – A term used to describe water where most of the chemicals that make water hard have been removed by chemical replacement.

Alkalinity – This is due to the presence of bicarbonate (HCO_3^-) or hydroxyl (OH^{-1}) ions. The pH is above 7. It is also called temporary hardness.

Acidity – This is due to the presence of hydrogen ion and is measured by a pH less than 7.

Deposits – Solids that buildup on metal surfaces inside the process. Heat exchangers are favorite targets because elevated surface temperatures permit evaporation to take place and leave solids behind. Scale is a name given to a typical deposit.

Corrosion –This as used here as a chemical attack on metal. Embrittlement, stress crack, and pitting are common corrosion mechanisms. Oxygen reacting with iron is an example.

Precipitation –As applied to water treatment is caused by the reaction of a soluble chemical with another chemical that produces an insoluble product (precipitant).

Coagulation –Is a process that neutralizes the negative charges on precipitants which allows small particles to collect into larger entities. Both suspended particles and particles formed by chemical reaction (precipitation) can be so small that physical removal is difficult unless coagulated into a size that will settle out or can be filtered out.

Flocculation –Is the bridging or sticking together of coagulants that form large agglomerates. Chemicals that encourage flocculation (called flocculating agents) are commonly used in water treatment.

Deaeration –Is the removal of gases from water. Oxygen is usually the target.

Clarification –Is the removal of suspended solids by chemical treatment to form larger particles (sludge or floc). A side benefit is removal of chemicals that cause temporary hardness.

Temporary Hardness –Is caused by the carbonate ion (OH^-) in natural water.

Permanent Hardness –Is caused by selected metal salts dissolved in natural water such as calcium and magnesium.

Total Hardness –Is the sum of temporary and permanent hardness and is measured as the chemical equivalent of calcium carbonate even though the actual chemicals may be sulfates, nitrates, or chlorides as well as carbonates. Hardness is usually reported as parts per million, ppm.

<u>External and Internal Water Treatment</u> –Terms used to define where the treatment takes place. External treatment is applied before the water is put into the process and is the focus of this course. Internal treatment takes place in the process. Internal treatment will be introduced in specific processes such as the water in a cooling tower operation.

<u>pH</u> –*A scale of 1 to 14 that is used to quantify acidity (hydrogen ion H^+) and alkalinity (hydroxyl ion OH^-) concentration. A pH of 7 is neutral ($H^+ = OH^-$). Below 7 is acidic and above 7 is basic.* pH is define as: $$pH = \frac{1}{Log(H^+)}$$

Therefore, each unit change, say from 6 to 7 or the reverse 7 to 6 is a tenfold decrease or increase in H^+ ion concentration.

<u>Turbidity and Color</u> –Used to describe opaque or unclear water. Light does not penetrate well. May contribute to odor also.

<u>Parts per million parts</u> –(Ppm) is a common way for reporting low concentrations. A laboratory may use milligrams per liter (1000 grams). Ppm can be converted to wt. % by dividing by one million and multiplying y 100.

$$\frac{1000\ ppm}{1,000,000} \times 100 = 0.1\ wt.\ \%$$

<u>Ion Exchange</u> –is a process that replaces an undesirable ion, say Ca^{++}, with a less harmful ion(H+).

<u>Sludge</u> –Solids produced by precipitation of impurities through use of soda-lime treatment.

CHAPTER 9

COOLING TOWER

INTRODUCTION

Surface water rarely reaches ambient temperature during the heat of the day. This fact that water can cool itself by contact with air has been known and studied for a long time. The general subject is called Psychrometry. Psychrometries covers the technology used in open recirculating cooling system, which include ponds and cooling towers.

Water keeps cool by the change of a small amount of liquid water into vapor (evaporation). This change in phase takes a large amount of heat energy per pound of water evaporated. The latent heat is supplied by the sensible heat of the liquid water so the temperature falls the main body of water that remains. What is of interest to industry is that a small amount of water is lost to evaporation with latent heat input, and a large amount of residual water is cooled by removal of sensible heat that is the source of the latent heat.

An elaborate facility is not necessary to cool water by evaporation. A pond with sufficient holdup times will do. If the pond water is aerated, the process is speeded up by better contact between air and water. However, a lot of stored water, a lot of surface area, and a lot of time are required to provide the coolant capacity of a typical petrochemical plant requires if natural evaporation is the sole heat removal mechanism in use. The advantages are low investment and cost.

The use of water once through that is pumped from rivers and wells is still a common practice. However, as the competition for water with municipalities increases on a yearly basis when a shortage eventually will occur, the municipality will be served

first. Other concerns with the use of surface water such as thermal and chemical pollution are very much in the public domain and cannot be overlooked as essential controls in acceptable plant operations.

In an increasing number of locations, the open circulating system that utilizes a cooling tower, has become a practical, cost effective process to provide cooling water (See Figure).

FIGURE 1

OPEN RECIRCULATING SYSTEM

COOLING TOWER

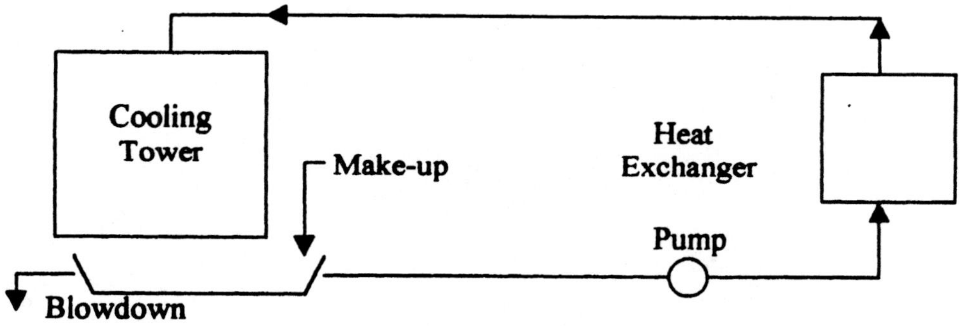

System	Characteristic	Disadvantages
Cooling tower	Average temperature Range 20° to 30°F	Corrosion Fouling Scale High air humidity

As a result, a multitude of devices under the general name of cooling towers have been invented and continuously refined to efficient force water to cool itself by evaporation. Cooling towers fall into two general categories: Forced Draft and Induced Draft (See Figure 2). The induced draft cooling tower will serve as a basis for the study of the technology involved.

In forced draft towers the fan is located on the air stream entering the tower. This tower is characterized by high air entrance velocities and low exit velocities, therefore, the towers, are susceptible to recirculation thus having a lower performance stability. The fans can also be subject to icing under conditions of low ambient temperature and high humidity. Usually, forced draft towers are equipped with centrifugal blower type fans which, although requiring considerably more horsepower than propeller type fans, have the advantage of being able to operate against the high static pressures associated with ductwork. Therefore, they can either be installed indoors (space permitting), or within a specially designed enclosure that provides significant separation between intake and discharge locations to minimize recirculation.

In induced draft towers, the fan is located on the air stream leaving the tower. This causes air exit velocities which are three to four times higher than their air entrance velocities. This improves the heat dispersion and reduces the potential for recirculation. Induced draft towers have an air discharge velocity of from 3 to 4 times higher than their air entrance velocity, with the entrance velocity approximating that of a 5 mph wind. Therefore, there is little or no tendency for a reduced pressure zone to be created at the air inlets by the action of the fan alone. The potential for recirculation on an induced draft tower is not self-initiating and, therefore, can be more easily quantified purely on the basis of ambient wind conditions. Location of the fan within the warm air stream provides excellent protection against the formation of ice on the mechanical components.

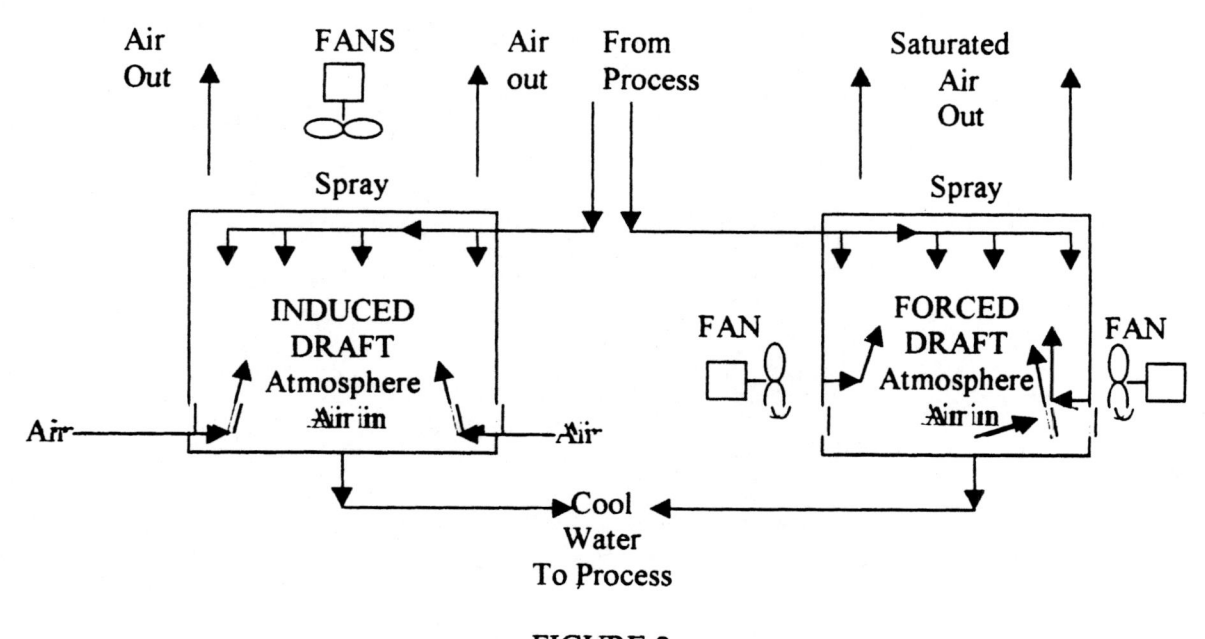

FIGURE 2

MOST COMMON COOLING TOWER CONFIGURATIONS

The atmosphere from which a cooling tower draws its supply of air incorporates infinitely variable psychrometric properties, and the tower reacts thermally or physically to each of those properties. The tower accelerates that air; passes it through a maze of structure and fill; heats it; expands it; saturates it with moisture by evaporation; scrubs it; compresses it; and responds to all of the thermal and aerodynamic effects that such treatment can produce. Finally, the cooling tower returns that "used up" stream of air to the atmosphere.

Cooling by evaporation is a natural occurrence and is a simple process to remove heat from liquid water by taking a small loss to evaporation. That is the principle of the cooling tower. However, there is a catch. The water vapor has to be dispersed into the air so there has to be a place for it. That place must come from an increase in humidity.

There is an upper limit on the water vapor air can hold called the dew point or saturation temperature. The amount of additional water air can accept is the difference in the water content at saturation and the amount of water already present in the air, that difference is defined as the relative humidity. Relative humidity determines the capacity of a cooling tower.

Cooling towers are a free standing industrial process. Each independent unit or tower is called a cell. A cell can be operated independently or in tandem with other cells. Therefore, the maximum cooling capacity of a tower installation is determined by the number of cells in service at any given time. Overall cooling capacity is increased by putting more cells on line.

The controlling factors in determining the capacity of a cooling tower are the relative humidity of the inlet AIR and the vapor pressure of the water that the air contacts (See Figure 3).

An understanding of the terms humidity, a term used to define the water content of air, is necessary in order to understand the technology that drives a cooling tower. Air has the potential to permanently hold water vapor. The maximum amount of water the air (humidity) can hold increases and decreases as the temperature of the air rises and falls. At 100% humidity, air contains all the water it can hold at a specified temperature. Technical descriptions of the maximum water content in air include saturated air or air at its dew point. The dew point of air is technically, the temperature where the water vapor content measured by the partial pressure of the water vapor in the air is equal to the vapor pressure of water at the same temperature. Any excess water will condense forming small droplets called dew or fog.

The driving force to humidify air is the vapor pressure of water. The vapor pressure is fixed by the temperature of the water. Until the pressure of water vapor in the air (called the partial pressure) reaches the saturation or dew point, the vapor pressure of the liquid water at any given temperature is driven to supply the shortfall by an increase in partial pressure by evaporation until the partial pressure equals the vapor pressure. Heat energy is required to convert the liquid water to vapor. This phase change from liquid to vapor requires simultaneous mass and energy transfer by evaporation.

The conversion of a liquid to a vapor consumes a large amount of latent heat energy. In fact, the amount of heat required to vaporize a pound of water would cool that same pound of water approximately 1000°F.

At any humidity less than 100%, the air can hold additional water vapor (See Figure 3). This means that the partial pressure of water vapor in the air is less than the vapor pressure of the water that is in contact with the air. Water continues to evaporate to satisfy the pressure difference, which is the driving force in the cooling tower.

In summary, the driving force in a cooling tower is the difference between the vapor pressure of liquid water and maximum partial pressure of water vapor that the air can hold at the air temperature at any point in the tower. The driving force that a cooling tower capacity can be determined from the dew point of inlet air.

FIGURE 3

RELATION OF HUMIDITY IN AIR

AND VAPOR PRESSURE OF WATER

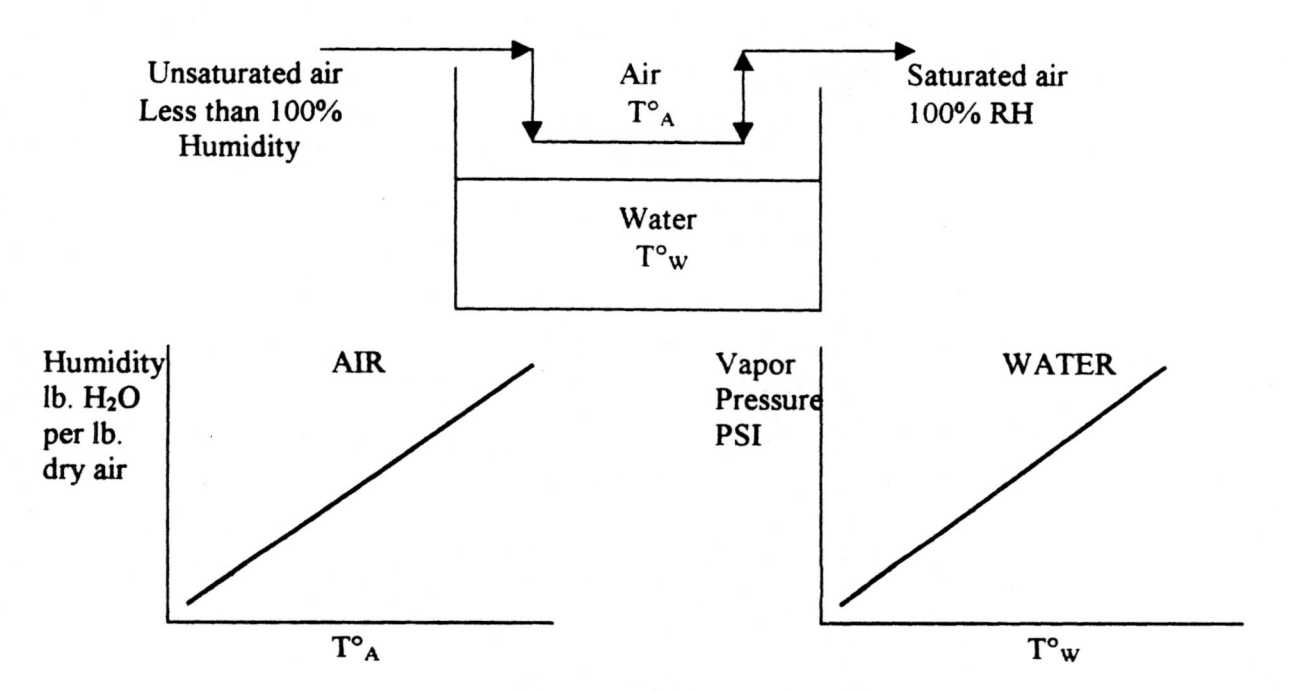

The technical problem is that air will not accept additional water vapor beyond its saturation level called the dew point or relative humidity (See table 1). At the dew point, which is also a relative humidity of 100%, condensation is as likely as evaporation with any decrease or increase in temperature.

The amount of water the air can take from the tower as vapor evaporated from the liquid is calculated as the humidity difference of the air in and out of the tower as pounds of water per pound of dry air. This can be visualized by comparing the water content of saturated air at two temperatures (See Table 1).

TABLE 1

ENTHALPY AND HUMIDITY

OF AIR

Air, Temperature, °F	Enthalpy, BTU/lb. of Dry Air	100% Relative Humidity, lb. Water per lb. Dry Air
30	10.92	0.0035
40	15.23	0.0052
50	20.30	0.0077
60	26.46	0.0111
70	34.09	0.0158
80	43.69	0.0223
90	55.94	0.0312
100	71.73	0.0432
110	92.33	0.0594
120	119.51	0.0815

THE COOLING TOWER AT WORK

Cool water is used to remove heat energy from an industrial process. A shell and tube heat exchanger would be typical.

The heated water, which contains the heat removed from the process, is returned to the top of a cell and sprayed (usually upward) to disperse it into small droplets in order to increase surface are (See Figure). This speeds up both heat and mass transfer through increased surface area that results in better contact between the air and water. The water falls down through the tower, which provides the necessary holdup time for heat and mass transfer to take place as the water evaporates. The vapor pressure of the water is the driving force behind the phase change. Evaporation is the process by which hot water, returning from the process, releases its heat to the atmospheric air is cooled and ready to return back to the process. The water vapor is retained in the air as an increase I humidity.

Air is forced upward through the tower to carry away the evaporated water. The amount of water evaporated is set by the increase in humidity. The velocity of the air is controlled by blowers and the cross sectional area of the tower. The blowers may pull (induced) or push (forced) the air flow through the tower. The maximum velocity is limited by an acceptable entrainment level of water droplets in the exit air.

The key point to remember is that the water is mostly cooled (90%) by the evaporation of a portion of itself and not by an increase in the air temperature (10%).

Time is required to transfer mass and energy. The holdup time of the water is set by tower height and free fall across slats called flu

Figure 4
Typical Cooling Tower

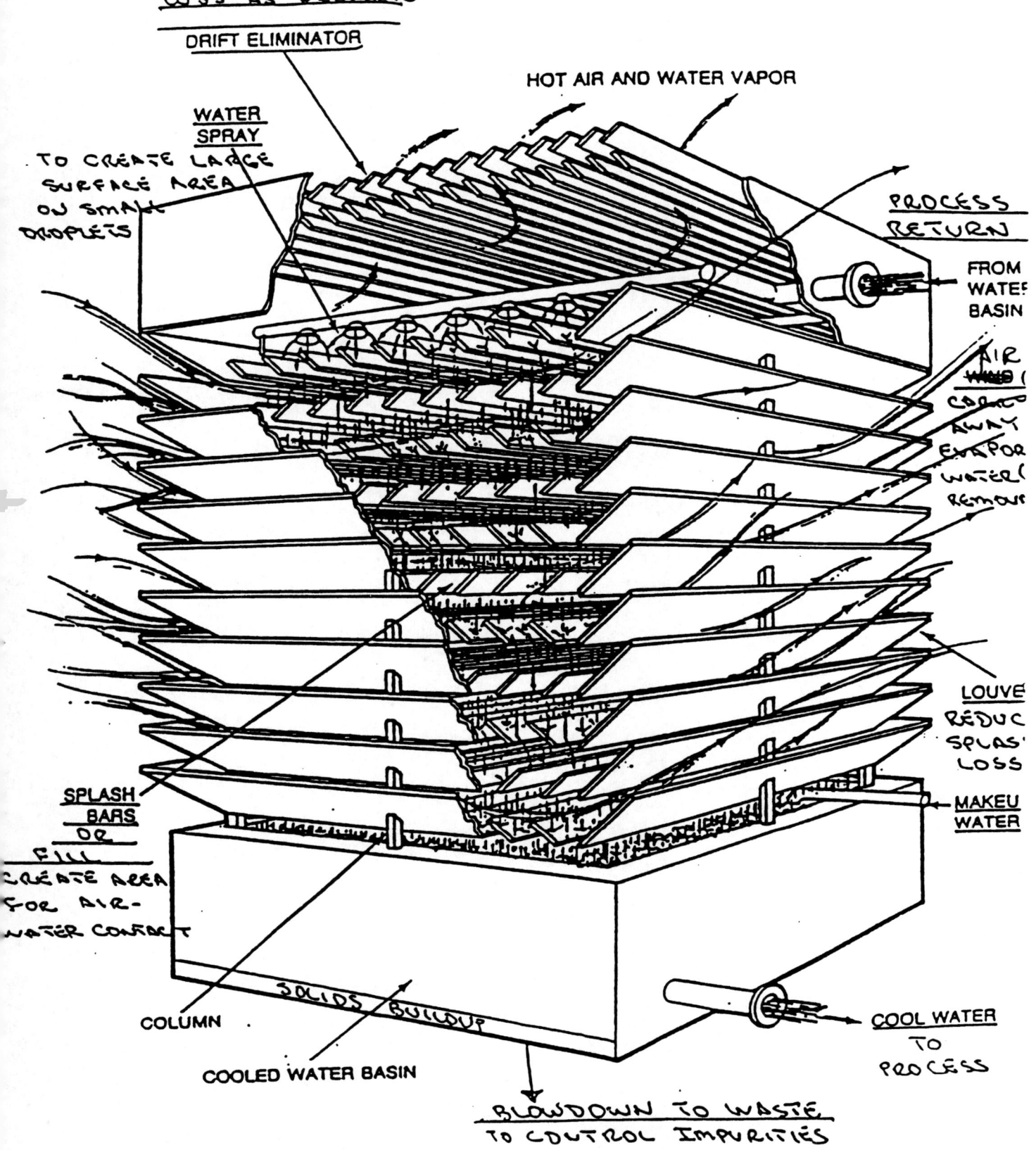

PTEC 2431   Industrial Processes

SECTION 7: Cooling Tower Operation

The maximum water evaporated = 0.0264 - 0.0092 = 0.0172 lb. per lb. of dry air.

Latent Heat needed = 1000 BTU/lb. x 0.0172 = 17.2 BTU

1000 BTU/lb. = heat vaporization of one pound of water.

17.2 BTU/lb. = heat of vaporization of 0.0172 lbs. of water. The 17.2 BTU is provided by the sensible heat in the water.

If it is assumed that the sensible heat of the dry air is negligible, the exit water temperature can be easily calculated since the latent heat to evaporate the water is provided by the sensible heat in the return water.

$$\Delta T = \frac{Q}{(wt.)(Cp)}$$

WT. = 1 pound
Cp = BTU/lb./°F = 1.0 for water

$$= \frac{17.2}{(1.0)(1.0)}$$

Range = 17.2

85°F - 17.2°F = 67.8°F   Which is the exit water temperature
85°F is the Inlet water temperature

Suppose that the relative humidity of the inlet air is 20%. Then, from the chart, the water content would be 0.0035 lbs./lb. dry air. The water evaporated lb.H_2O/lb. dry air.

= 0.0264 - 0.0035 lb. H_2O/lb. dry air

= 0.0229 lb. H_2O/lb. dry air

As seen, $\Delta T = \frac{0}{(wt.)(Cp)}$

$$= \frac{22.9}{(1.0)(1.0)}$$
= 22.9°F

A POUND OF WATER THROUGH A COOLING TOWER

A typical cooling tower could have heated water returned from the process at 85°F. The atmospheric air to the tower could be at 75°F and a 50% relative humidity. The two temperatures and the relative humidity are supplied by standard instrumentation.

With this information, and using the Psychrometric Chart in Figure 2, we find that the inlet air contains 0.0092 lb. of water per pound of dry air.

When the 85°F water is sprayed into the top of the tower, the air coming up from the bottom is assumed to be saturated (100% humidity) at 85°F. From Figure 2, we find that saturated air at 85°F holds 0.0264 lb. H_2O/lb. dry air.

The maximum water evaporated = 0.0264 - 0.0092 = 0.0172 lb. per lb. of dry air.

Latent Heat needed = 1000 BTU/lb. x 0.0172 lb. = 17.2 BTU/lb.

(1000 BTU/lb. = heat of vaporization of one pound of water)

17.2 BTU/lb. = heat of vaporization of 0.172 lbs. of water. The 17.2 BTU is provided by the sensible heat in the water.

If it is assumed that the sensible heat of the dry air is negligible, the exit water temperature can be easily calculated since the latent heat to evaporate the water is provided by the sensible heat in the return water.

$$\Delta T = \frac{Q}{(wt.)(Cp)} \quad \text{Wt.} = 1 \text{ pound}$$
$$Cp = BTU/lb./°F = 1.0 \text{ for water}$$

$$= \frac{17.2}{(1.0)(1.0)}$$

$$= 17.2°F$$

85°F − 17.2°F = 67.8°F   Exit water temperature
85°F is the Inlet water temperature

Suppose that the relative humidity of the inlet air is 20%. Then, from the chart, the water content would be 0.0035 lbs./lb. dry air. The water evaporated lb. H_2O/lb. dry air.

$$= 0.0264 - 0.0035 \text{ lb. } H_2O/\text{lb. dry air}$$

$$= 0.0229 \text{ lb. } H_2O/\text{lb. dry air}$$

Latent heat needed = 1000 BTU/lb. x 0.0229 = 22.9 BTU/lb.

As seen, $\Delta T = \dfrac{Q}{(\text{wt.})(C_p)}$

$$= \dfrac{22.9}{(1.0)(1.0)} = 22.9°F$$

Suppose we use 1.5 lbs. of dry air per lb. of water. Keep the inlet air at 75° and 50% humidity.

Water In = (1.5)(0.0092) = 0.0138 lbs. H_2O/lb. dry air

Water Out = (1.5)(0.0264) = 0.0396 lbs. H_2O/lb. dry air

Water Evaporated = 0.0396 − 0.0138 lb. H_2O

$$= 0.0258 \text{ lbs. } H_2O$$

ΔT = 25.8°F (calculate as shown above)

T_{out} = 85.0°F − 25.8°F

= 59.2°F Exit temperature of the water.

That is a respectable approach to the technical limit at 55°F dew point.

COMMON COOLING TOWER PROBLEMS

Common physical problems due to impurity concentration in cooling towers fall into four categories.

- Corrosion
- Scale
- Fouling
- Microbiological contamination

The most common indications of the excessive presence of one or more of the problems:

1. Loss of heat transfer
2. Reduced equipment life and failure
3. Increased maintenance requirements

Impurities in cooling water buildup through mechanisms.

1. Each 10°F drop in the water temperature across the cooling tower, results from a 1% loss of cooling water due to evaporation. When this 1% loss of water occurs, there is a 1% increase in the concentration of dissolved solids in the water. If the cooling water is allowed to continue concentrating without any control, eventually the mineral solids will get so high that scale deposits will occur.

2. The impurities in the makeup water needed to replace water lost to evaporation, entrainment, and blowdown buildup in the system.

3. If the objective were to scrub the air to remove dust and other impurities, a cooling tower would do an excellent job. Also, impurities washed from the air concentrate in the system due to evaporation. As a means of controlling the impurity level, a purge

of water called blowdown is taken. The purpose is to reduce fouling of the heat transfer surfaces in the process heat exchangers serviced by the tower water. Each time the increase in the total amount of solids in the tower is equal to the amount of solids already present in the makeup water, this is equal to 1 cycle of concentration in the tower. For example, if the makeup water dissolved solids are 500 ppm and the cooling water dissolved solids are 1000, then the cycles of concentration is 2.

$$\frac{C_A \text{ in circulating water}}{C_A \text{ in makeup water}} = \text{cycles of concentration}$$

C_A is concentration of impurity A in the circulating water and in the makeup water.

BLOWDOWN

To prevent the dissolved solids concentration in the cooling water from becoming so high that mineral scale deposits begin to form, part of the recirculating water is deliberately and continuously dumped from the system. This procedure is called blowdown or bleed-off. This procedure controls the build-up of dissolved solids by replacing the more highly concentrated system water with an equal volume of fresh, less concentrated makeup water.

Blowdown, or bleed-off, is the continuous removal of a portion of the water from the circulating system. Blowdown is used to prevent the dissolved solids from concentrating to the point where they will form scale. The amount of blowdown required depends upon the cooling range (the difference between he hot and cold water temperature) and the composition of the make-up water (water added to the system to compensate for losses by blowdown, evaporation, and drift).

Blowdown to some extent can be limited by proper chemical treatment, which allows operation at higher impurity concentrations.

RECIRCULATION

For an induced draft tower operating under calm conditions, with a vertically rising plume, entering and ambient wet-bulb temperatures can be considered to be equal. This important technical point is not valid if some portion of the saturated air leaving the tower being induced back into the inlet air.

This undesirable situation is called "recirculation." The potential for recirculation is primarily related to wind force and direction, with recirculation tending to increase as wind velocity increases. For that reason, accepted codes under which cooling towers are tested for thermal performance limit wind velocity during the test to 10 mph.

FIGURE 4

RECIRCULATION

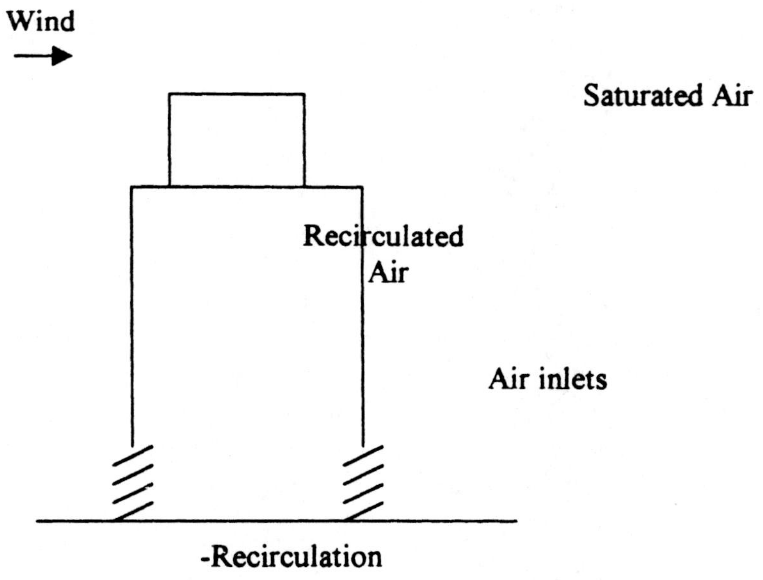

FOUR MAJOR PROPERTIES OF COOLING WATER

The corrosion and scale potential of cooling water can be monitored through four measurable properties.

- Conductivity
- Hardness
- Alkalinity
- pH

Conductivity is the measure of how well water will "conduct" an electrical current. Pure water, with no dissolved minerals in it, will not conduct an electrical current. As minerals in the water accumulate, the conductivity increases. Conductivity is a direct measurement of the amount of dissolved solids in the water. As the conductivity of water increases the potential for corrosion and scale formation increases.

Calcium and magnesium are the components that make up water "hardness". Calcium and magnesium make water "hard to wash with". Hardness minerals react with soap and make it necessary to use more soap.

Alkalinity is one of the most critical components in water. If the alkalinity is too high, scaling deposits can form. If the alkalinity is too low, corrosion can result. The two forms of alkalinity that are important in cooling water are: carbonate and bicarbonate alkalinity. Under certain conditions, calcium and carbonate can react together to form calcium carbonate, sometimes called a "lime deposit".

pH is an important indication of corrosive potential. As a rule, acidic water is more corrosive but can be controlled with chemical treatment.

CHEMICAL ATTACK

Many chemicals are corrosive, foul heat transfer surfaces, and foster microbiological growth (See Table).

TABLE

COOLING SYSTEM WATER IMPURITIES

CHEMICAL	PROBLEM	CORRECTION
Aluminum	Fouling	Dispersant
Amine	Microbiological fouling/corrosion	Biocide/surfactant/inhibitor
Ammonia	Microbiological fouling/corrosion (especially copper)	Biocide/surfactant/inhibitor
BOD	Microbiological fouling	Biocide/surfactant
Calcium	Fouling	Dispersant
Chlorides	Corrosion (especially stainless steel)	Inhibitor
Conductivity, ppmhos	Corrosion	Inhibitor
Copper	Corrosion	Inhibitor(azole)
Cyanides	Corrosion	Inhibitor
Disulfide	Corrosion	Inhibitor
Esters	Microbiological fouling	Biocide/surfactant
Ether	Microbiological fouling	Biocide/surfactant
Fluoride	Fouling	Dispersant
Hydrocarbon –total	Microbiological fouling/fouling	Biocide/surfactant
Hydrocarbon –light	Microbiological fouling	Nonoxidizing biocide/surfactant
Hydrocarbon –heavy	Microbiological fouling/fouling	Biocide/surfactant
Hydrogen sulfide	Corrosion/microbiological fouling	Inhibitor/nonoxidizing biocide/surfactant
Iron	Fouling	Dispersant
Methanol	Microbiological fouling	Biocide/surfactant
Nickel	Fouling	Dispersant
Oil and grease	Microbiological fouling/fouling	Biocide/surfactant
pH	Corrosion/fouling	pH control
Sodium	See conductivity	
Suspended solids	Fouling	Dispersant
Vanadium	Fouling	Dispersant

CORROSION

Corrosion is the mechanism by which processed metals are reverted back to their natural oxidized state. In nature, metals are mixed with oxygen. For example, iron ore is iron mixed with oxygen. In the process of making steel the oxygen is removed to get to the pure iron. This processed steel tends to rust, to react with oxygen and return to its natural state.

The affect of corrosion can be summarized as follows:

- Corrosion destroys cooling system metal
- Corrosion product deposits in heat exchangers
- Heat transfer efficiency is reduced by deposits
- Leaks in equipment develop
- Process side and water side contamination occurs
- Water usage increases
- Maintenance and cleaning frequency increases
- Equipment must be repaired and/or replaced
- Unscheduled shutdown of plant

CAUSES OF CORROSION

In general, for every 18°F increase in water temperature the chemical reaction rate doubles. So, as the cooling water temperature increases, the corrosion rate increases.

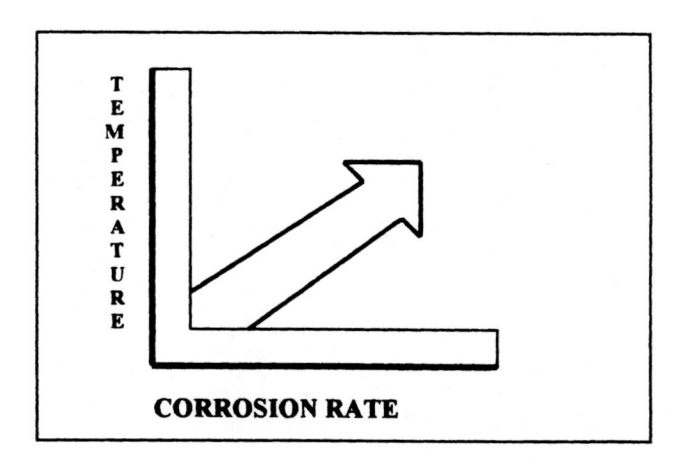

Temperature and Corrosion

It is the dissolved solids in the cooling water that complete the electrical circuit, from the cathode back to the anode. In general, the higher the dissolved solids, the higher the conductivity, the higher the corrosion rate.

The corrosion rate is dramatically affected by the cooling water pH. As the pH drops, the corrosion rate increases. A low pH allows reactions at the cathode to accelerate.

CORROSION PREVENTION

As we stated earlier, corrosion cannot be completely stopped. However, there are several things you can do to help control or minimize corrosion. The following methods can be used to help control corrosion.

Use corrosion resistant alloys. Once the cooling system is built this becomes an expensive alternative.

Adjust system pH. Increasing pH will reduce corrosion rate but will also increase the scaling potential

Apply protective coatings. These are generally not very effective because of the rough metal surfaces and difficulty of maintaining the integrity of the coating. Pre-treating system metallurgy with materials such as Nalprep after mechanical or chemical cleaning has been very effective at minimizing corrosion during system start-up.

Apply chemical corrosion inhibitor programs.

CORROSION CONTROL

The metals utilized in a cooling tower are susceptible to corrosion in varying degrees. This is often true of even the most sophisticated metals, although they can usually withstand deeper excursions into the realm of corrosion, and for longer periods of time, than can the more "standard" metals. Circulating water having corrosion characteristics beyond those anticipated in the tower's design requires treatment. This may be due to high oxygen content, carbon dioxide, low pH, or the contact of dissimilar metals. Where correction of the source of trouble cannot readily be made, various treatment compounds may be used as inhibitors, which act to build and maintain a protective film on the metal parts.

Since most water system corrosion occurs as a result of electrolytic action, an increase in the dissolved solids increases the conductivity and the corrosion potential. This is particularly true of the chloride and sulfate ions. Therefore, blowdown is a very useful tool in the fight against corrosion.

FOULING

Fouling is the accumulation of solid material, other than scale, in a way that hampers the operation of equipment or contributes to its deterioration.

The following sources can produce these fouling materials in a cooling water system:

SOURCE	FOULANT
• Makeup Water	• Silt, Sand, Mud, and Iron
• Air through Cooling Tower	• Dust and Dirt
• Internal Contaminants	• Process Contaminants, • Oils, Corrosion Products, • Microbiological Growth

Each of these materials are suspended solids. They have a tendency to stock together and eventually settle out of the water. When this happens they form a deposit called scale on metal surfaces in the cooling water system, which interferes with the flow of water and the transfer of heat form the process.

SCALE

Scale tends to form more readily in the hot areas of the system. This includes heat transfer zones and low flow areas of the cooling system. Heat exchangers with the cooling water on the shell side are the most vulnerable to fouling problems on the outside surface of the tubes.

Anywhere scale forms into deposits, an ideal area for localized pitting type corrosion is in place. Corrosive bacteria can also thrive under these deposits. Pitting can eventually penetrate the tube wall and start a leak.

The affects of the formation of deposits on metal surfaces summarized as:

- Decreased heat transfer by lower overall coefficients
- Loss of system capacity
- Increased maintenance requirements
- Higher chemical inhibitors demand

The most fundamental operation strategy to control scale is to control the purity and source of the makeup water. Other useful techniques is a filter to continuously remove solids and careful attention to blowdown. Close adherence to chemical treatment practices recommended by the tower designer as a basis for his design as also important.

MINERAL SCALE

Mineral scale forms in the hot parts of the cooling system, which is usually in the heat exchangers. The build up of deposit reduces the ability of the water to take away heat from the process. (Recall film resistances in heat transfer through a wall.) An insidious side effect is pitting type corrosion under the scale deposit. This localized attack will eventually and unpredictably eat through the metal piping of the cooling system and start a leak. The major effect of scale formation is decreased heat transfer capability and loss of capacity primarily in process equipment.

The following factors influence mineral (metal salts) scale formation.

- Concentration of minerals usually controlled by blowdown and makeup water quality.
- Water temperature is usually a built-in design parameter. Additional installed capacity to operate within design parameters is a solution.

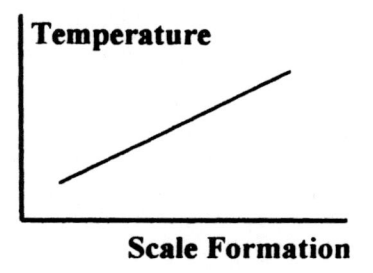

```
Temperature
   |          /
   |        /
   |      /
   |    /
   |  /
   |/_____
         Scale Formation
```

- Water pH is controlled by chemical treatment. Either excessive alkalinity or acidity is a poor operating condition.

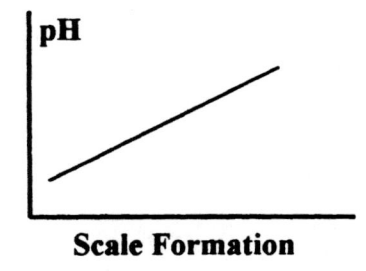

```
pH
   |          /
   |        /
   |      /
   |    /
   |  /
   |/_____
         Scale Formation
```

Note that this increase in pH that assists scale formation usually deters corrosion.

- Suspended solids build-up increases scale formation
- Low water velocity (dead zones) increase scale formation

<u>Chemical Treatment</u>

The principle scale-forming ingredient in cooling water is calcium carbonate, which has a solubility of about 15 ppm and is formed by the thermal decomposition of calcium bicarbonate. The maximum amount of calcium bicarbonate that can be held in solution depends upon the temperature and the free carbon dioxide content of the water. Raising the temperature or reducing the free carbon dioxide, at the point of equilibrium, will result in the deposition of scale.

Sulfuric acid or one of the polyphosphates is most generally used to control calcium carbonate scale. Various proprietary materials containing chromates, phosphates or other compounds are available for corrosion control.

MICROBIOLOGICAL GROWTH

Your cooling system is the ideal environment for microscopic organisms to live and breed. They enter your system from the air or makeup water and find everything they need to survive there. Left uncontrolled they multiply quickly and can completely foul a cooling system in a very short period of time.

There are three kinds of microorganisms that are troublesome to cooling water systems; Bacteria, Algae, and Fungi. All three have different nutritional requirements and grow in different parts of the cooling system.

BACTERIA

Bacteria are the most dangerous microorganisms in cooling water because they can cause the most damage. There are four classifications of bacteria. All four thrive in cooling water but because of their differing needs they live in different parts of the system.

ALGAE

Algae must have sunlight to survive and grow. That is why it is generally found only on cooling tower decks or other open surfaced areas. Algae grows from a microscopic size cell into a large mass called an "algae mat". These growths can plug water flow on top of the cooling tower, plug screens and foul equipment. Even after algae is dead, it can provide shelter for anaerobic corrosive bacteria and food for other organisms. Algae also consumes oxidizing biocides.

FUNGI

Fungi are a group of microorganisms that use the carbon in wood fibers for food. This carbon is found in the cellulose fibers that give wood its strength. The lumber found

in cooling towers is the prime target for fungal attack. Fungi cause either surface or internal rotting of the wood. Fungi are capable of collapsing an entire cooling tower as a result of deep wood rot.

WHAT CAUSES MICROBIOLOGICAL GROWTH

Recirculating cooling system is the ideal environment for microbiological growth.

Present at all times are:

- Nutrients (NU3, Oil, Organics)
- Temperature (75°F - 135°F)
- pH (6.0 –8.0)
- Location (light/no light)
- Environment (aerobic/anaerobic)

MICROBIOLOGICAL GROWTH PREVENTION

System Design Considerations

There are several considerations in system design that can help deter the growth of microorganisms.

- Clean sludge from tower basins regularly
- Use plastic fill and mist eliminators to reduce problems with fungi
- Cover tower decks to inhibit algae growth

Chemical treatment with Biocides

A good chemical program is critical to controlling microorganisms. Three general classifications of chemicals are used for the control of microorganisms:

Oxidizing biocides

Non-oxidizing biocides

Biodispersants

Oxidizing biocides work by penetrating the microorganism's cell wall and "oxidizing" or "burning up" the internals of the organism. No microorganism is resistant to this oxidation process. Oxidizers, at sufficient levels, kill any microorganism they contact.

When added to cooling water, oxidizing biocides react with certain contaminants present in the water. The most common contaminants are ammonia and hydrocarbons, as well as, microorganisms. These contaminants affect the performance of the oxidizing biocide because they consume biocide. The amount of contaminants that react with the oxidant is called the "oxidant demand." The amount of oxidant that is unreacted or left over is called the "free available oxidant." It is the presence of free available oxidant that

provides satisfactory microbio control. It is necessary to maintain a prescribed level of free available oxidant for good control.

One of the limitations of oxidizing biocides is that they only kill the microorganisms they contact. Microorganisms that grow under deposits or inside wood supports in cooling towers will not be affected by oxidizing biocides.

There are many types of non-oxidizing biocides. Each one has a specific mechanism of kill. Many of them act by interfering with the metabolism of the microorganism, or destroying the microorganism's cell wall. They are typically fed at high dosages, in single slugs. Unlike oxidizing biocides, they do not generate a residual oxidant level.

Non-oxidizing biocides must be fed at a relatively high dosage rate to be lethal. Typically they are too expensive to use as a stand-alone biocide program except in smaller systems. Large systems may use non-oxidizing biocides in emergencies, for example, during process leaks.

COOLING TOWER

TERMS AND NOMENCLATURE

<u>Air Inlet</u> –Opening in a cooling tower through which air enters. Sometimes referred to as the louvered face on induced draft towers.

<u>Air Rate, G</u> –Mass flow of dry air per square foot of cross-sectional area in the tower's heat transfer region per hour. [lb./(sq. ft. hr)]

<u>Air Velocity, V</u> –Velocity of air-vapor mixture through a specific region of the tower (i.e. the fan). (ft./min.)

<u>Ambient Wet-Bulb Temperature, AWB</u> –The wet-bulb temperature of the air encompassing a cooling tower, not including any temperature contribution by the tower itself. Generally measured upwind of a tower, in a number of locations sufficient to account for all extraneous sources of heat. (°F)

<u>Approach</u> –In practice, a cooling tower never reaches the lowest temperature that it is possible to cool the water, which is the wet-bulb temperature. The difference between the actual temperature of the cooled water and the wet-bulb temperature is called the approach.

<u>Basin</u> –The holding tank for the cooled water. Located at the bottom of the tower.

<u>Time per cycle</u>

<u>Blowdown</u> –The amount of the circulating water stream that is purged to control impurities.

<u>Blower</u> –A squirrel-cage (centrifugal) type fan; usually applied for operation at higher-than-normal static pressures.

BTU(British Thermal Unit) –The amount of heat gain (or loss) required to raise (or lower) the temperature of one pound of water 1°F.

Capacity –the amount of water (gpm) that a cooling tower will cool through a specified range, at a specified approach and wet-bulb temperature. Unit: gpm

Cell –Smallest tower subdivision which can function as an independent unit with regard to air and water flow; it is bounded by either exterior walls or partition walls. Each cell may have one or more fans and one or more distribution systems.

Circulating Water Rate –Quantity of hot water entering the cooling tower. Unit: gpm.

Cold Water Temperature –Temperature of the water leaving the collection basin, exclusive of any temperature effects incurred by the addition of make-up and/or the removal of blowdown. Unit: °F. symbol: CW.

Dew Point Temperature –This has the same meaning as 100% relative humidity and the adiabatic saturation temperature. At constant pressure, it is the temperature where: Evaporation and Condensation occur at the same rate in a closed system (no mass or heat energy in or out). The water content of the air remains at its maximum concentration.

Drift –circulating water loss from the tower as liquid droplets entrained in the exhaust air stream. Units percent of circulating water rate or gpm. [For more precise work, an L/G parameter is used, and drift becomes pounds of water per million pounds of exhaust air (ppmw).]

Drift Eliminators –An assembly of baffles or labyrinth passages through which the air passes prior to its exit from the tower, for the purpose of removing entrained water droplets from the exhaust air.

Dry-Bulb Temperature, DB –The temperature of the entering or ambient air adjacent to the cooling tower as measured with a dry-bulb thermometer. (°F)

Evaporation Loss –Water evaporated from the circulating water into the air stream in the cooling process. (per cent of circulating water rate or gpm)

Distribution System –Those parts of a tower, beginning with the inlet connection, which distribute the hot circulating water within the tower to the points where it contacts the air for effective cooling. May include headers, laterals, branch arms, nozzles, distribution basins, and flow-regulating devices.

Fill –The term used to describe the spashboards or other means to force the water to take a longer path down the tower than normal free fall would provide.

Forced Draft –Refers to the movement of air under pressure through a cooling tower. Fans of forced draft towers are located at the air inlets to "force" air through the tower.

Heat Capacity, Cp –The amount of heat energy needed to raise the temperature of one lb. of a substance one degree. Stated as BTU/lb./°F or °C.

Heat Load –Total heat to be removed from the circulating water by the cooling tower per unit time. Units: BTU per min. or BTU per hr.

Height –On cooling towers erected over a concrete basin, height is measured from the elevation of the basin curb. "Nominal" heights are usually measured to the fan deck elevation, not including the height of the fan cylinder. Heights for towers on which a wood, steel, or plastic basin is included within the manufacturer's scope of supply are generally measured form the lowermost point of the basin, and are usually overall of the tower. Unit: ft.

Hot Water Temperature –Temperature of circulating water entering the cooling tower's distribution system. Unit: °F. Symbol: HW.

Induced Draft –Refers to the movement of air through a cooling tower by means of an induced partial vacuum. Fans of induced draft towers are located at the air discharges to "draw" air through the tower.

Inlet Wet-Bulb Temperature –See "Entering Wet-Bulb Temperature."

Holding Capacity –The total amount of water held by the cooling water system as expressed in gallons.

Latent Heat –The amount of heat energy required to vaporize (evaporate) a liquid. Usually stated as BTU/lb. or calories/gram.

Leaving Wet-Bulb Temperature –Wet-bulb temperature of the air discharged from a cooling tower. Unit: °F. Symbol: LWB.

Psychrometric Chart

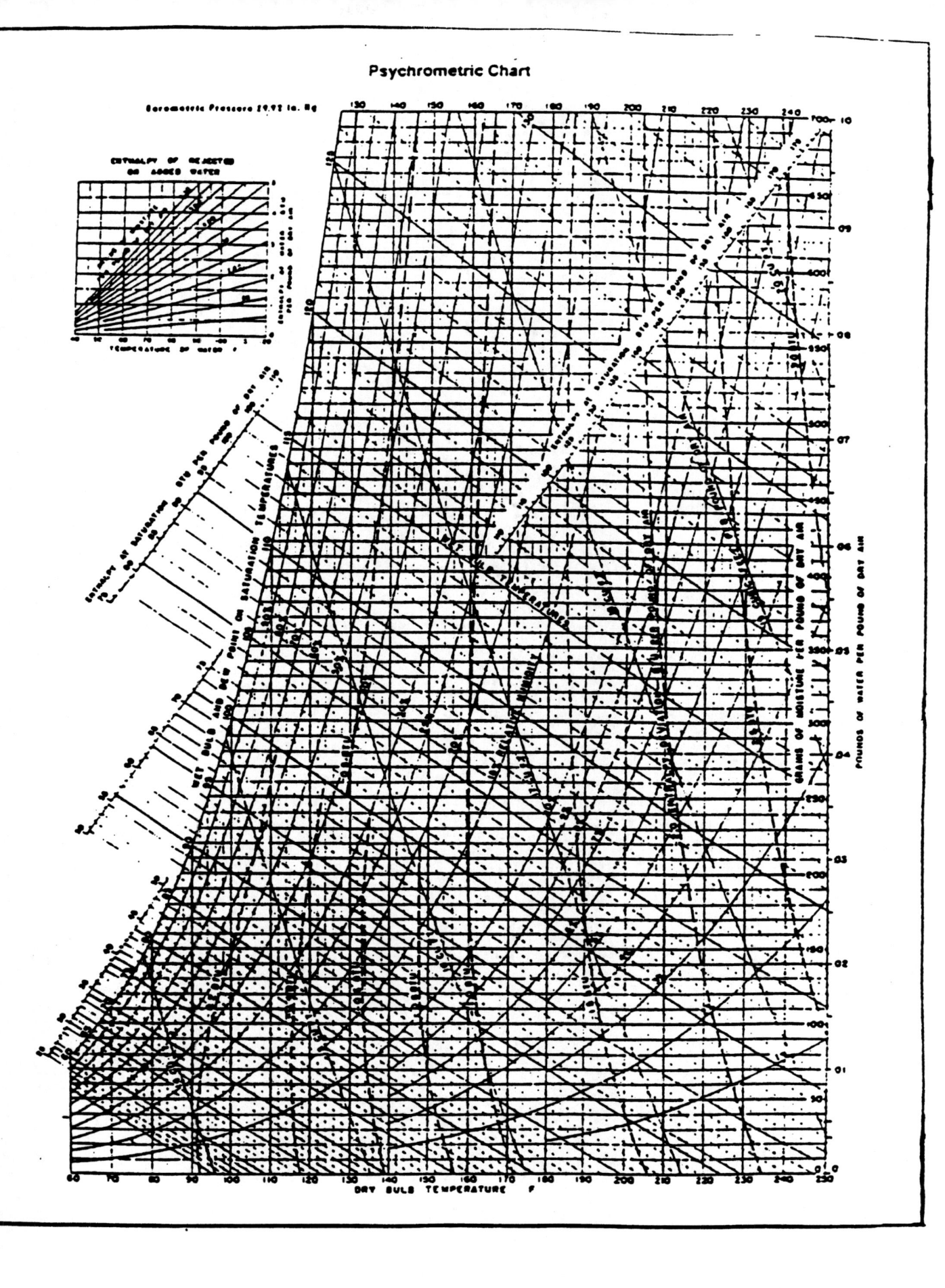

BASIC PROCESS TECHNOLOGY

SECTION 10: Extraction   T. D. Felder

I. INTRODUCTION

Extraction is the separation of a component from a liquid solution or mixture by contact with a liquid solvent that is immiscible with the solution or mixture. One or more components in the solution are also soluble in the solvent. It is essentially a mass transfer of a material between liquid phases.

The extraction works best when one or more of the components are more soluble in the solvent than in the mixture. The driving force in extraction then is the solubility difference of components between the mixture and the solvent. In essence, a component in a mixture is transferred to the solvent in amounts proportional to its mutual solubility between the mixture and in the solvent. The transfer is also proportional to the mass ratio between the two phases called the lever principle. (SEE FIGURE 1)

Extraction is normally used where distillation is not practical. A typical case would be separation of a solid inorganic component that is present in an inorganic—organic chemical mixture. Extraction is also very effective in removing polar compounds like inorganic acids from non-polar components by water extraction.

Extraction can be used to recover a useful product or to remove an impurity from a useful product.

II. PRINCIPLES

The driving forces in extraction are mutual solubility and the mass ratio which is called the lever principle. The lever principle is used to calculate the expected separations efficiency in a mixture where the solubilities of the component to be removed is known.

Take a mixture of A and B and remove A by extraction with a solvent S. The solubility of A in S is known. The concentration and solubility of A in B is known. The solvent S is immiscible with B and has a reasonable solubility for A. Also, there must be a density difference between S and B, which permits gravity separation by decantation or settling of the two phases when extraction is complete.

Hopefully, A is highly soluble in S but it need not be infinitely so. We can use a large amount of S which "levers" A into S.

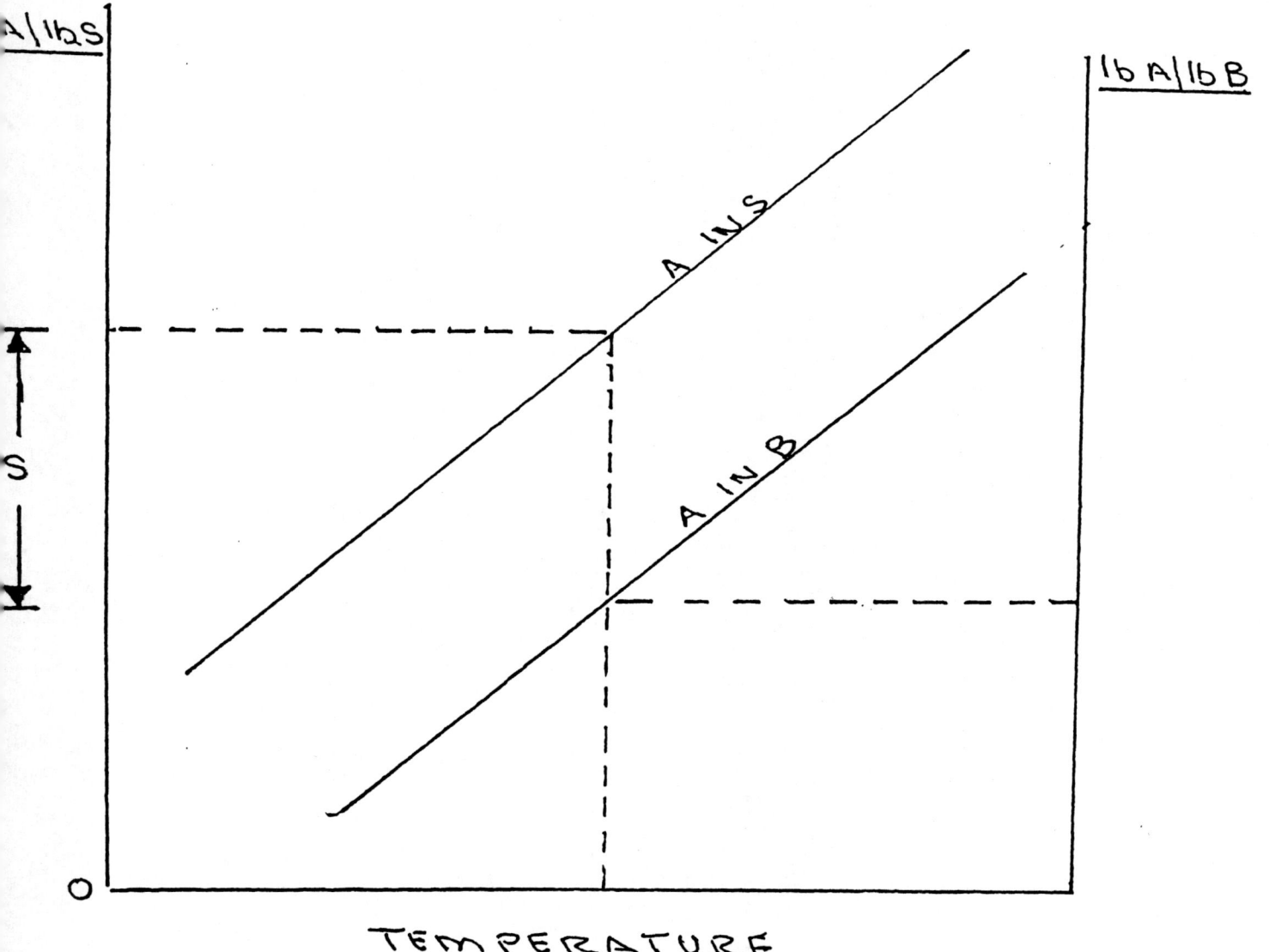

TYPICAL SOLUBILITY CURVES

FIGURE 1

SEPARATOR
(DECANTER)

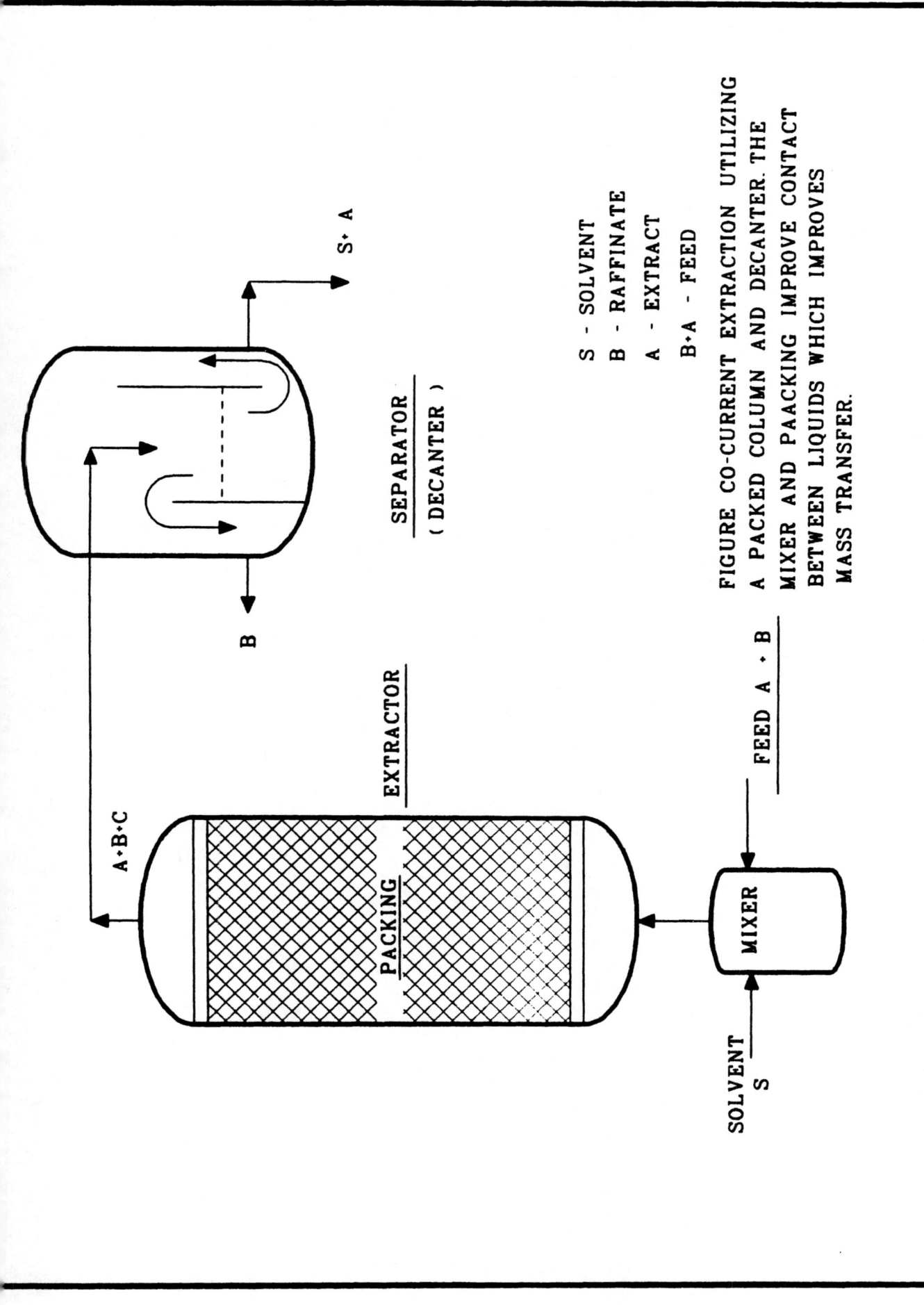

EXTRACTOR

PACKING

FEED A + B

SOLVENT S

- S - SOLVENT
- B - RAFFINATE
- A - EXTRACT
- B+A - FEED

FIGURE CO-CURRENT EXTRACTION UTILIZING A PACKED COLUMN AND DECANTER. THE MIXER AND PAACKING IMPROVE CONTACT BETWEEN LIQUIDS WHICH IMPROVES MASS TRANSFER.

CO-CURRENT EXTRACTION | SCALE:N.T.S. | S. DEL RIO | 11/12/99

BASIC PROCESS TECHNOLOGY

SECTION 10 Extraction — T. D. Felder

The use of multiple contacts of pure S with the A and B mixture to extract A is very effective and is most often used in industry. (See Figure: 3.) Multiple contacts make extraction useful when dealing with very low concentrations of A in B. It is essentially the same principle that makes a number of small rinses more effective in washing a surface clean than one large amount of rinse.

The process does not involve significant heat energy in most cases so thermal instability of A or B is not a problem and heat removal is not required. Decantation (settling) is a very simple and predictable way to use density difference to separate B from S + A.

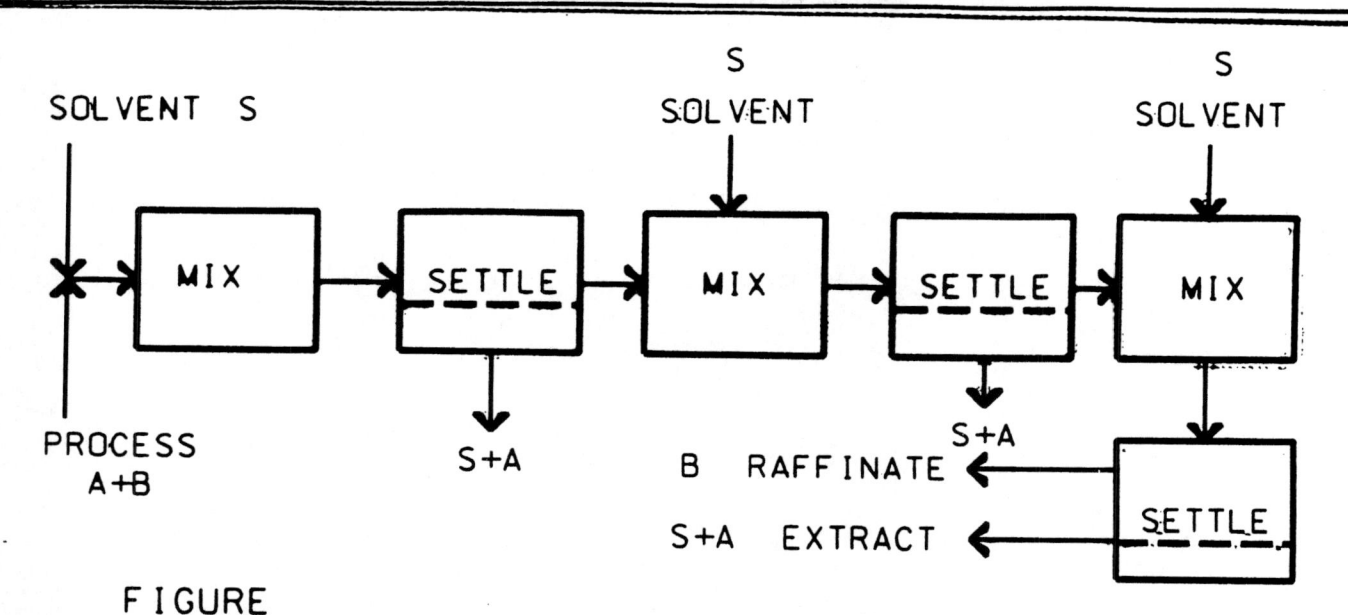

FIGURE
BLOCK DIAGRAM OF <u>MULTI-STAGE CO-CURRENT EXTRACTION</u>
MUTIPLE INJECTIONS OF SOLVENT INTO FEED WHICH OPTIMIZES SOLVENT CONSUMPTION

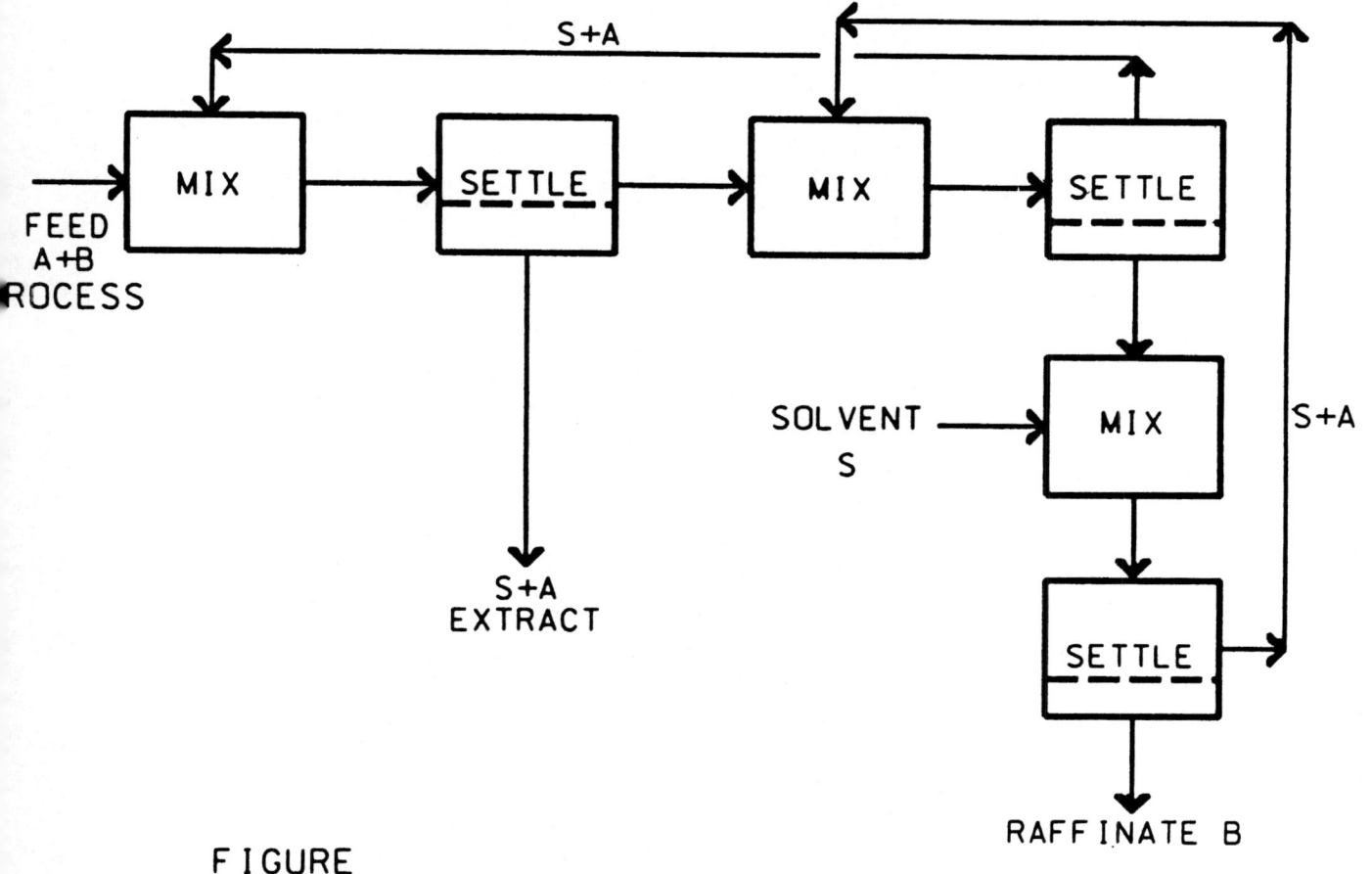

FIGURE
BLOCK DIAGRAM OF <u>MULTI-STAGE COUNTER-CURRENT EXTRACTION</u>
MULTIPLE CONTACTS OF SOLVENT AND FEED MAXIMIZES CONCENTRATION OF A IN SOLVENT

THE VICTORIA COLLEGE | 11/24/99

BASIC PROCESS TECHNOLOGY

SECTION 10 Extraction T. D. Felder

III. TERMS AND SYMBOLS

<u>Extraction</u> is the separation of a mixture by treatment with a selective solvent specific to one or more components.

<u>Solvent</u> is a liquid that is immiscible with one component to be separated as it dissolves another.

<u>Heat Sensitive</u> refers to a material that degrades, chemically reacts, or in any way loses useful properties when heated.

<u>Feed</u> is the mixture sent to an extraction process in order to be separated.

<u>Extract or Extract Layer</u> is the solvent with the component to be separated dissolved in it.

<u>Raffinate</u> is the mixture left behind after a component(s) has been extracted.

<u>Miscible</u> describes liquid components that form uniform mixtures when brought together. Individual components lose their identity.

<u>Immiscible</u> describes liquid components that will not mix and will separate into two phases if a density difference exists. Each component retains its identity such as color.

<u>Stage</u> is one cycle of mixing and solvent extraction.

<u>Co-current</u> is an operating mode where all flow is in the same direction through the process.

<u>Countercurrent</u> is an operating mode where two streams flow in opposite directions through the process.

<u>Mixer</u> is a device to efficiently mix two substances.

<u>Settler</u> or decanter is an equipment piece that allows time for immiscible liquids to separate by gravity into two phases.

<u>Extractor</u> is an equipment piece that provides time and contact for a solvent to dissolve a component from a mixture.

<u>Emulsion</u> is a state where two normally immisicible liquids are mixed so intimately that the small particles do not separate upon standing. It can be a major problem in extraction efficiency.

BASIC PROCESS TECHNOLOGY

SECTION 12: Extraction T. D. Felder

IV. HOW IT WORKS TECHNICALLY (See Figure 1,2, and 3)

The extraction process needs two conditions in place for control of the separation by mass transfer.

1. There must be intimate contact between the solvent and the mixture from which a component is to be extracted. Good contact is accomplished by thorough mixing which creates large component surface areas, which is important in effective mass transfer. The mixing is usually carried out, mechanically, by propeller, turbine or simply a pump.

   Fifty or sixty pipe diameters of components under turbulent flow conditions is a fairly good pipe mixer for low viscosity fluids and is simple to operate.

   Viscosity of either the solvent or the mixture to be extracted is a concern in mixing. Emulsion formation is always a possibility so excessive shear in the mixer must be avoided.

2. Time is needed for the mass transfer to take place. The time needed is usually proportional to the concentration and solubility of the extract, and the ratio of solvent to extract. This lever principle is useful in predicting the number of stages needed to complete a desired separation.

After extraction, the extract and raffinate layers are separated. This process, called settling, works well in a simple decanter or settler for immiscible fluids with a good density differential. For fluids with a small density differential, or where a high decree of separation is required, a centrifuge is often used. In counter current operation, the components separate naturally in the extractor.

The component that has been extracted into S is separated by distillation (liquid) or evaporation (solids). S is recycled for reuse in the extractor. The component may be a product, intermediate or waste.

The raffinate can be the product and sent on for further refining.

BASIC PROCESS TECHNOLOGY

SECTION 10: Decantation T.D. Felder

I. INTRODUCTION
Decantation is one of the most useful and simple ways to separate two immiscible liquids. It is not effective in producing pure products but can reduce the load on more precise separations such as distillation or extraction.

A decanter is also useful as an environmental protection device to contain or collect process spills and separate them from casual water. Building sumps serve this purpose. Very large operations usually have a building decanter to separate contaminants from surface water such as rain.

II. PRINCIPLES (Figure 1)
The decanter is effective in separation of immiscible liquids but there must be a density difference. Decantation works best when the density difference between the liquids to be separated is large and the mutual solubility is small. It will not separate suspensions and emulsions which are stable mixtures of normally immiscible liquids.

The separation time required in the decanter to produce two phases is mostly determined by the density difference. The higher the density difference the lower the holdup time (vessel size) is required to make the separation at a constant feed rate.

The liquids being separated will continue to contain dissolved amounts of each other up to their mutual solubility which may be very small but the physical separation is never perfect.

III. SYMBOLS AND TERMS
<u>Decanter or settler</u> is a vessel that provides sufficient time for two or more immiscible liquids to separate into two phases and effect a separation. It provides an upper and lower drawoff point for the two phases.

<u>Interface</u> is the separation point between the high density and low density phases in the decanter.

<u>Baffles</u> are internal dividers that allow under flow (dense phase) and overflow (light phase) into separate chambers.

<u>Light Phase</u> (layer) is the less dense material that rises to the top.

<u>Heavy Phase</u> (layer) is the more dense material that settle to the bottom.

km: Rev. 3 5/00

BASIC PROCESS TECHNOLOGY

SECTION 13: Decantation
T. D. Felder

IV. HOW IT WORKS

A decanter or settler operates on the density difference between two immiscible liquids. In petrochemical applications, the separation is usually between water and a hydrocarbon. It is often used to complete another process such as extraction.

The key is the immiscibility or lack of mutual solubility between the liquids. The separation is not highly efficient for several reasons.
1. There is always some degree of mutual solubility.
2. The degree of separation depends upon the settling time which is finite in an economically sized vessel.
3. There is always some degree of mixing due to liquid movement inside the decanter.
4. Any emulsion or suspension formed will not separate without treatment.

There are empirical equations that predict settling time versus separation efficiency that relate time, droplet size, density gradient (difference) and solubility. For good performance data, experimental results are preferred.

Decanter Components (See Figure 1)

The decanter is normally a tank that is divided into three sections by two baffles. One baffle (1) separates a section of the decanter by a seal at the bottom but allows overflow at the top.

The second baffle (2) is open at the bottom for underflow but provides a seal at the top which allows underflow.

The largest section of the vessel (3) provides the holdup time for separation of the feed to take place under non-turbulent conditions.

When the separation is complete, an interface forms between the light and heavy phases. The interface is controlled along with the feed rate to obtain the desirable separation. The drawoff rate of upper level make and the lower level tails is used to control the interface. A cloudy or indistinct interface is often a symptom of emulsion or suspension problems due to excessive mixing.

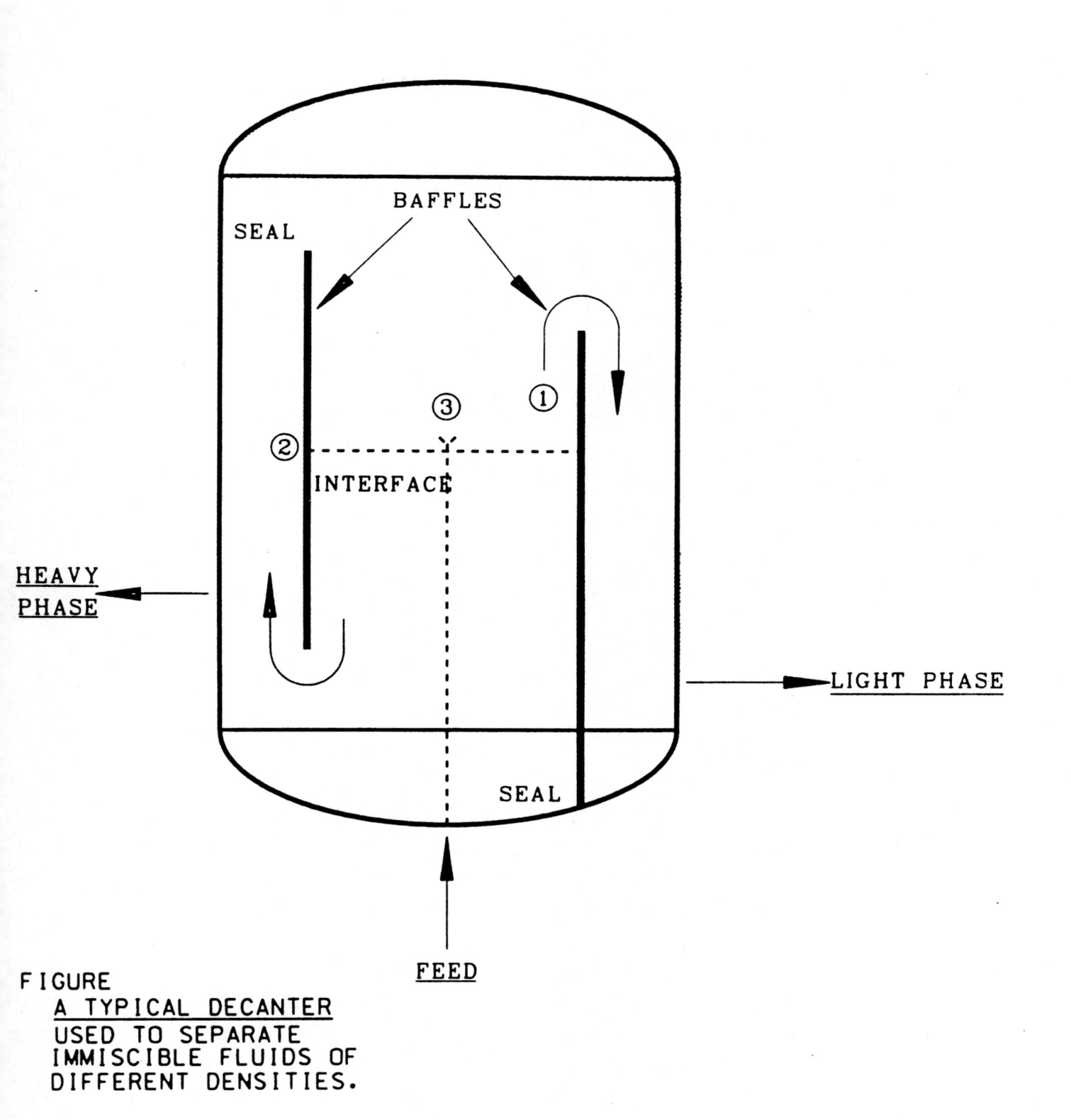

FIGURE
A TYPICAL DECANTER USED TO SEPARATE IMMISCIBLE FLUIDS OF DIFFERENT DENSITIES.

BASIC PROCESS TECHNOLOGY

SECTION 14: Crystallization   T.D. Felder

I. INTRODUCTION

Crystallization is used to separate a solid from its mother liquor (solvent). The driving force to form crystals is solubility, which is controlled normally by temperature. Some materials are so soluble that crystallization is impractical, but a sister process called evaporation is effective.

For many high molecular weight organics, crystallization is a practical choice of separation. One large advantage is that an excellent separation from impurities often occurs since crystal formation is capable of producing some of nature's purest substances.

The crystallization process is one of the most precisely controlled of all the petrochemical processes and the most difficult to operate. The major operating objectives are uniform crystal size and crystal purity. Crystals that are too large or too small result in process problems in filtration, purity, plugging, bulk density and caking.

When properly designed and operated, crystallization produces products of a purity level rarely attained by any commercial process. But to maximize yield yet effectively separate impurities, a very precise control of temperature and concentration is essential.

II. PRINCIPLES

Reducing solubility by lowering the temperature of a solution is the driving force that causes crystals to form and grow. The mass of crystals formed can be predicted from a solubility curve (see Figure 1).

As shown in Figure 1, the solubility of $MgSO_4$ increases with an increase in temperature and decreases with a fall in temperature. This is perfectly normal. A decrease in solubility to below the saturation temperature causes crystals to form (crystallization) in an amount equal to the reduction in solubility.

As you can see from Figure 1, there are three hydrates or molecular forms of magnesium sulfate. $MgSO_4 \cdot 7H_2O$, the most useful one called Epson Salts, precipitates between 60 and 120 °F. The point made here is that the precipitation temperature can be as important as the solubility in producing the desired product from a solution.

It is also true that the size of the crystals for any fixed amount of precipitate is a function of the number of nuclei (new crystals) and existing crystals present in the crystallizer for crystals to grow on. At temperatures close to saturation, the growth on crystals already formed is much higher than the formation of nuclei. This characteristic of crystal formation permits crystal size control.

PTEC 2431 INDUSTRIAL PROCESSES
SECTION 11.4 : CRYSTALLIZATION   T.D FELDER

FIGURE 1
SOLUBILITY OF $MgSO_4$

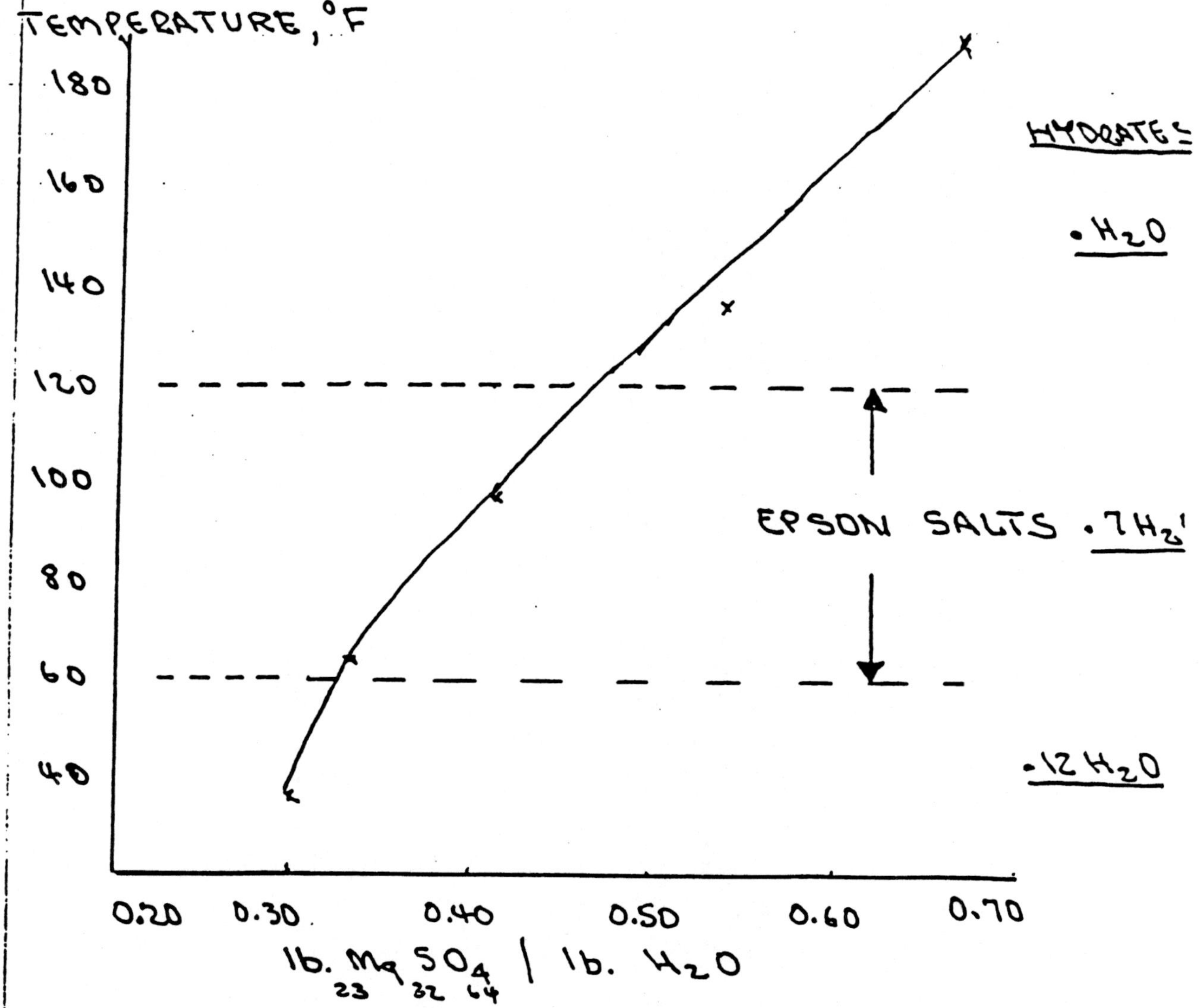

Temperature, °F vs lb. $MgSO_4$ / lb. H_2O

Hydrates:
- $\cdot H_2O$ (above ~120°F)
- Epson salts $\cdot 7H_2O$ (between ~60°F and ~120°F)
- $\cdot 12H_2O$ (below ~60°F)

WT. FRACT.	lb. $MgSO_4$ PER lb. H_2O	TEMP. °F
0.23	30	37
0.25	33	62
0.30	43	98
0.35	54	135
0.40	67	190

BASIC PROCESS TECHNOLOGY

SECTION 14: Crystallization					T.D. Felder

Crystal number and therefore crystal size must be controlled for good quality and processibility. The one key is to control the number of nuclei formed. This is difficult in an untreated solution. One treatment is to add small crystals to the solution. This process called seeding is essentially the recycle of small crystals in the mother liquor after filtration and allowing them to grow to the desired size in the crystallizer by providing adequate holdup time at a specified temperature. The approach is often used in batch crystallization.

If high crystal yield is to be achieved crystal growth must take place at a low concentration differential between the saturation temperature and the actual temperature inside the crystallizer. In effect, driving force is low (ΔC), growth rate is low, so hold-up time must be relatively long to produce an adequate yield. As a consequence, crystallizers are large vessels per pound of product produced.

III. TERMS AND SYMBOLS

<u>Crystallizer</u> is a device that provides time, temperature and solubility control that is favorable to crystal formation.

<u>Crystals</u> are solid particles with precise forms and dimensions grown on nuclei precipitated from solution or on seed crystals in the solution.

<u>Mother Liquor</u> is the solvent plus dissolved solid (solute) from which crystals are grown and separated.

<u>Magma</u> is the crystal and mother liquor mixture.

<u>Solute</u> is the material dissolved in the solvent.

<u>Solubility</u> is the pounds of solute that will dissolve in a pound of solvent at the saturation temperature.

<u>Solubility Curve</u> is the plot of the effect of temperature on solubility at saturation.

<u>Saturation point</u> is the temperature where the solvent is saturated with solute.

<u>Super saturation</u> is a condition where the temperature falls below the saturation point and no nuclei are formed.

<u>Nuclei</u> are the initial small particles that come out of solution.

<u>Hold-up time</u> is the volume of the vessel divided by the volume of the feed rate at constant level.

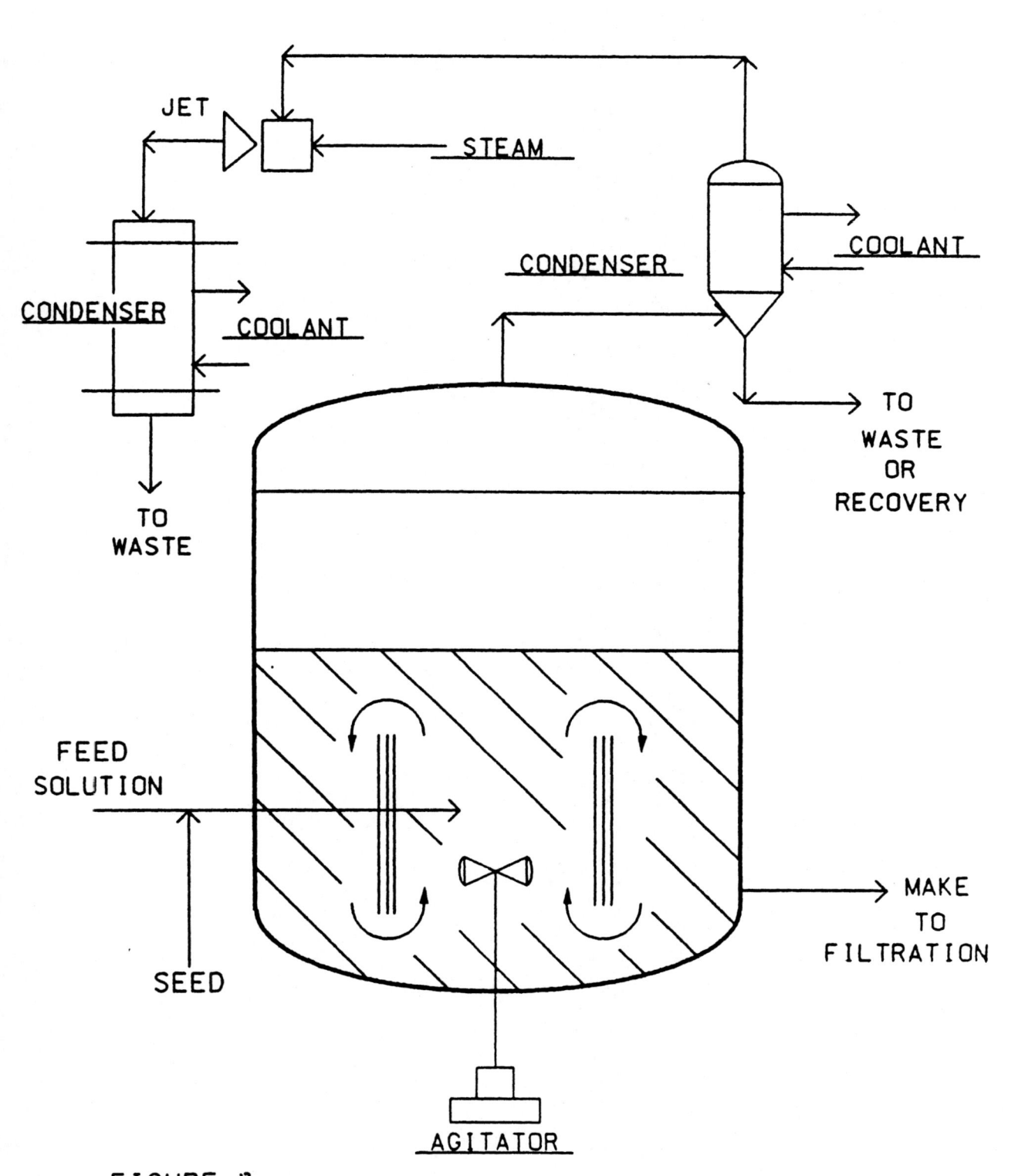

FIGURE 2
CONTINUOUS VACUUM CRYSTALIZER COOLING IS OBTAINED BY EVAPORATION OF A SMALL AMOUNT OF SOLVENT WITH LATENT HEAT PROVIDED BY SENSIBLE HEAT

| THE VICTORIA COLLEGE | VANESSA A. URESTE | 11/24/99 | DRAW 8 |

BASIC PROCESS TECHNOLOGY

SECTION 14: Crystallization   T.D. Felder

IV. HOW IT WORKS (Figures 2 and 3)

Crystallization is in principle a simple process. One takes a saturated solution and cools it below the saturation temperature. Solids precipitate and are separated as product by filtration.

However, crystallization to produce a high purity product of uniform crystal size is a very complicated process that utilizes close control of a number of variables. In this course the broad effects in a continuous crystallization system are used to illustrate principles.

Feed temperature and solubility
The temperature of the solution (fed) to the crystallizer has fixed the maximum concentration of the solute in the solvent. In practice the feed is superheated to prevent premature solids formation in the feed system. The solute concentration in the feed is close to the saturation point but always below it. In appearance, the solution is clear or at most a little cloudy.

Crystallization or operating temperature
The crystallization temperature is the temperature of the contents of the crystallizer. At this temperature, below the feed temperature solubility is reduced. The solution (mother liquor) is close to saturation and crystal growth is driven by a low concentration difference. Desired growth is primarily on existing crystals with little nuclei formation.

The solution is agitated or circulated to get good contact between crystal and solution which is necessary to achieve good mass transfer.

The amount of crystallization is closely controlled to maximize yield but leave soluble impurities behind.

This temperature is controlled in two ways:
1. The crystallizer can be operated under vacuum. Solvent is evaporated (flashed) which concentrates the solution and reduces solubility. The latent heat comes from the sensible heat of the solution, which has a cooling effect which further reduces solubility.
2. The crystallizer content is cooled by circulation through an external heat exchanger. The problem is potential physical damage to fragile crystals.

Operating Pressure
In a vacuum crystallizer, which uses evaporation to provide cooling, the pressure sets the boiling point (flash point) which in turn sets the saturation solubility temperature and controls the concentration in the mother liquor.

BASIC PROCESS TECHNOLOGY

SECTION 12: Chemical Reactions   T.D. Felder

I.   Introduction

II. Principles

III. How They Work

IV. Terms and Symbols

BASIC PROCESS TECHNOLOGY

I. INTRODUCTION

Chemical reactions that produce useful products are the justification for petrochemical plants. It takes specific and consistent standard operating procedures to ensure that reactions are safe, competitive and profitable.

There are four areas where good operations are essential for a healthy business:
- optimization of yield to salable products (productivity)
- control of energy release (safety)
- high utility to meet production goals (capacity)
- maintain product quality to meet customer needs (competitive)

A complete study of reaction kinetics is beyond this course, but we will cover the areas most important in operation of a chemical reaction.

Variable	Reaction Mode
concentration of reactants	polymerization
residence time (holdup)	oxidation
temperature	chlorination
contact mixing	cracking\reforming
pressure	hydrogenation
catalysis	neutralization
initiation	isomerization (later)
inhibition of reaction rate	

Reaction conditions are often mandated by solubility, vapor pressure, latent and sensible heat transfer and other physical properties of the reactants and products.

Many plants can produce a range of products by use of common reaction equipment operated with small changes in feedstock and reaction conditions. This is especially true in polymerization reactions. What is important to the Process Operator is how to use a set of slightly different operating conditions to produce a modified product. Even small errors in reactor conditions can result in unacceptable product quality.

Some reactions produce only one product but changes in reaction conditions are made for each production rate to optimize yield. Reaction conditions are continuously being studied to produce the desired product, at the optimum rate and quality.

In Table 1 are listed some of the variables important to pre-reaction, reaction and post-reaction operations.

REACTANTS

PREHEATER

HEAT OR COOL

CATALYST
OPTIONAL

PREHEATER

PRODUCTS

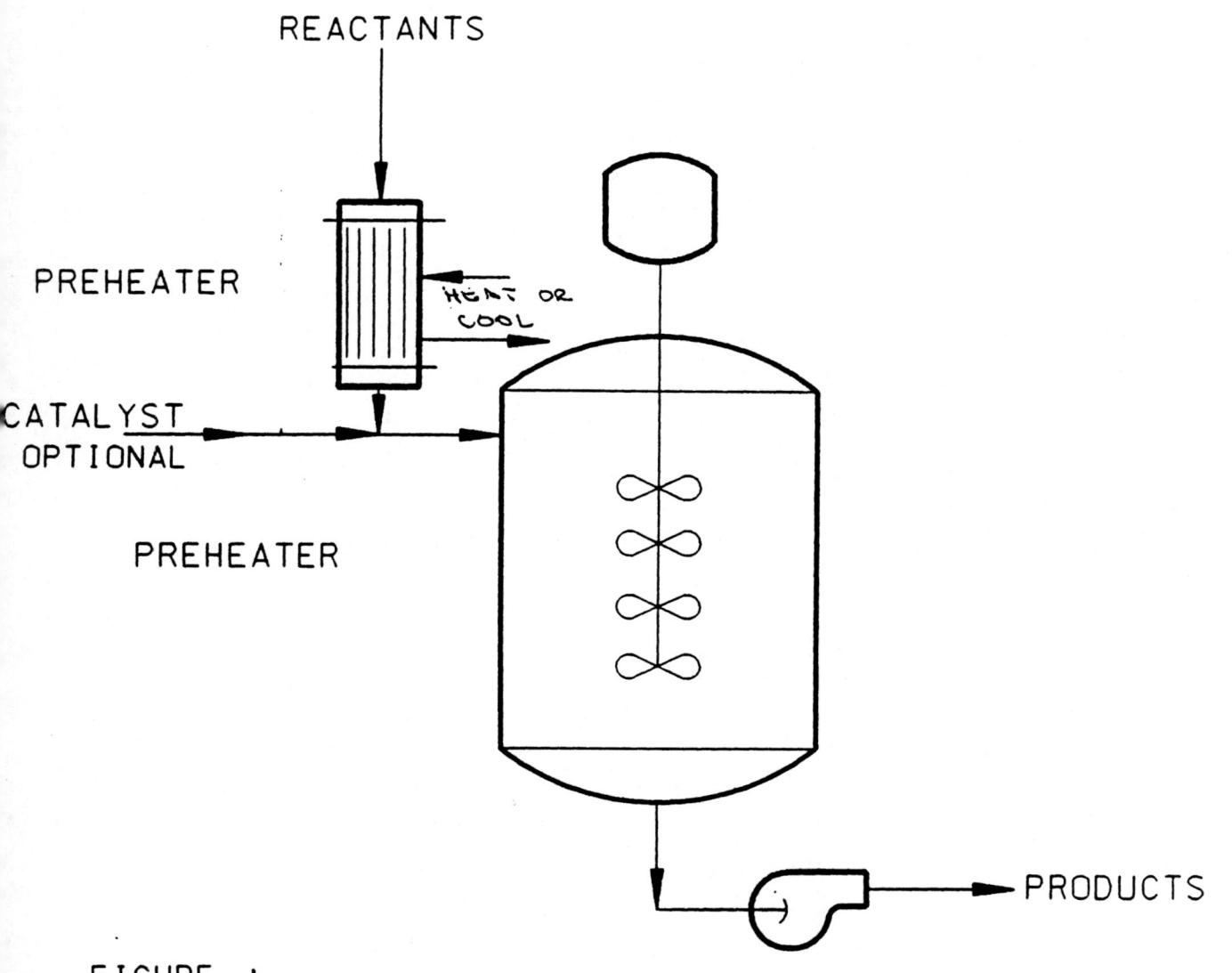

FIGURE 1

THE ADIABATIC REACTOR IS ONE OF
THE MOST VERSATILE IN INDUSTRY
THE AGITATOR MAINTAINS CONSTANT
CONGENTRATIONS AND TEMPERATURE
THROUGHOUT THE CONTENTS.

THE VICTORIA COLLEGE	ADIABATIC REACTOR-STIRRED
	FIGURE 1 11/22/99

SECTION 12: Chemical Reactions

Process Technology

T.D. Felder

TABLE 1
Typical Reaction Variables

PRE-REACTION	REACTION	POST-REACTION
Feed Preperation	**Facilities**	**Products**
drying	reactor	catalyst kill
mixing	pre-heater	
	circulation/mixing	secondary react (clean-up)
Catalyst Prep	heat removal	
solution		
Feed Stock	**Variables**	
Purification	temperature	temperature
consistency	pressure	hold-up time
	conversion	
	feed rate	
	hold-up time	
	Component contact	
	catalyst	

II. PRINCIPLES (FIGURE 1)

Yield

Yield is the most common factor used to measure performance of a reaction process because it directly impacts on rate and profitability. In order to better establish and maintain high operating standards, yield is measured and followed in two ways.

Overall Yield = $\dfrac{\text{Pounds of salable product shipped}}{\text{Pound of reactants}}$

Chemical Yield = $\dfrac{\text{Pounds of salable product exit reactor}}{\text{Pound of reactants}}$

Ideally, the ratio in both cases would be 1 to 1. Working to sustain and improve yield is always a major priority.

In order to develop yield improvement programs, one defines where yield is being lost. Overall yield loss commonly occurs in three ways. All three are very strongly affected by good or poor operating technique.

BASIC PROCESS TECHNOLOGY

Chemical Yield Loss = $\dfrac{\text{Pounds of useless products}}{\text{Pound of reactor feed}}$

Refining Yield Loss = $\dfrac{\text{Pounds of product to waste}}{\text{Pound of product refined}}$

Quality Yield Loss = $\dfrac{\text{Pounds of Off - Quality Product Produced}}{\text{Pound of quality product}}$

The off-quality product can be reworked in some cases. It is still a serious, once through yield loss, and an additional cost penalty to process off quality product in order to make it salable.

Reaction Rate

The reaction rate can be defined at any point in time using the principle of the Law of Mass Action.

For a Uni-Molecular Reaction at Constant Temperature

$$A \rightarrow B \text{ or } BC$$

$$k = \dfrac{2.303}{t} \text{ LOG } \dfrac{C_A}{C_{AO}}$$

k = velocity constant
t = lapsed time
C_A = amount of A that has not reacted
C_{AO} = initial concentration of A

A number of concentration units can be used for a and x as long as they are consistent.

The bi-molecular reaction is more important in industry and is used for discussion in this course. Constant temperature is assumed for simplification.

$$A + B \rightarrow C + D \rightarrow E-\text{etc.}$$

$$k = \dfrac{1}{t\,C_{AO}} \left[\dfrac{C_{AO} - C_A}{C_A} \right]$$

k = velocity constant
t = lapsed time
C_{AO} = initial concentration
C_A = amount of A that is unreacted

BASIC PROCESS TECHNOLOGY

The first key principle in the bi-molecular reaction above is that the reaction rate is proportional to the concentrations of reacted and unreacted material at constant temperature. The second principle is that reaction rate decreases at a linear rate with lapsed time.

The second way to relate time and rate is to state that rate decreases with time and reactants concentration due to conversion of reactants to products.

Notice that the rate can be effected by the concentration of each reactant component. This principle allows the use of an excess of one component to increase the rate and insure the total conversion of another component. This is usually an economic decision to best utilize the most expensive reactant.

A more practical equation relates <u>reactor volume</u> to time and concentration. Since most industrial reactors operate continuously, this equation is used as the basis to establish holdup time needed for a given production rate. Reactor volume determines holdup time at any given feed rate

$$k = \frac{2.303}{t} \text{LOG} \frac{C_o}{C_t}$$

k = rate constant lbs\FT³\HR
t = lapsed time
C_o = original concentration in number of moles per FT³ of reactor volume (consumption)
C_t = concentration of unreacted C after time t in number of moles per FT³ of reactor volume.

Activation Energy

Reactions require energy to get started and it is usually in the form of heat. This variable is called the activation energy and is defined by the Arrhenius Equation

$$A = \left(\frac{d \, Lnk}{dt}\right)(RT^2)$$

dLnk = change in reaction rate on the logarithmic scale
T = absolute temperature
R = gas constant
A = activation energy
dt = change in time

This equation holds true for multiple or simultaneous reactions. The principle here is that the activation energy is a function of the square of the absolute temperature. This leads to the general observation that reaction control depends upon control of the temperature.

Reaction Variables

The variables in a chemical reaction are highly interactive. A change in any major variable usually results in a major change in other variables and in the reaction outcome. However, each variable does exert more influence in some specific reaction area such as yield or rate. In this discussion, one variable is changed and the result studied.

Conversion is the amount of reactants converted to product divided by the amount of reactants in the feed. It is usually reported in %.

BASIC PROCESS TECHNOLOGY

1. % conversion for any reactant = $\dfrac{\text{Final concentration of reactant}}{\text{Original concentration of reactant}} \times 100$

2. % conversion for all reactants = $\dfrac{\text{lbs. reactants in - lbs reactants out}}{\text{lbs. reactant in}} \times 100$

1. Best fits batch reactions, and 2. best fits continuous reactions.

It is important not to confuse conversion with yield. Simply, conversion is a measure of <u>total</u> formed products while yield is a measure of <u>useful</u> products produced.

The major controls on conversion are temperature and holdup time.

Temperature

Temperature determines the rate of a chemical reaction at any fixed concentration of reactants. As the temperature increases, molecules move more rapidly, make contact more frequently and as a result react at a higher rate. It is mostly a probability factor in that more active molecules are more likely to meet.

In some chemical reactions, the rate can double with each 10° C increase in temperature. The importance of this generalization is that an out-of-control, exothermic reaction is a possibility with loss of temperature control that usually means an inability to remove heat energy at the rate that it is being generated. A chemical explosion is essentially an out of control, exothermic reaction.

The effect of temperature can be expressed by this equation:

$$Ln \dfrac{k_2}{k_1} = \dfrac{(A)}{(R)} \left(\dfrac{1}{T_1} - \dfrac{1}{T_2} \right)$$

A and R are constants
T_2 is the higher temperature (absolute)
T_1 is the initial temperature (absolute)
k_1 is the initial rate
k_2 is the higher rate

What the equation states in principle is that the reaction rate increases logarithmically (exponentially) with a temperature increase.

Temperature control is a major factor in the overall control system on a reactor.

Pressure

The effect of pressure on a chemical reaction is very predictable but is difficult to quantify. In liquid phase reactions, the rate can be related to vapor pressure. That is based on the known principle that the higher the vapor pressure the more active the movement of molecules. This results in more molecular contact and increases the possibility of a reaction taking place.

BASIC PROCESS TECHNOLOGY

In a vapor phase reaction, pressure increases the density of the reactants. This increases molecular concentration per unit volume which increases molecular collisions and the reaction rate. Again it is easy to predict this broad effect, but difficult to quantify.

One problem is that gases do not act ideally at elevated pressure and temperature. This is taken into account by use of a measure of non-ideality called an activation constant.

Concentration of Reactants

The Law of Mass Action states that the reaction rate is proportional to the logarithm of the initial and current concentrations of reactants at any point in time. Stated another way, the reaction rate decreases exponentially as the reactants are consumed or products appear. In reality, reactors must operate at a low concentration of reactants in the product in order to achieve a practical conversion rate.

Time

In principle, time is the easiest variable to control in a chemical reaction. The longer the reactants are in contact, the higher the conversion to products. The conversion slows down exponentially as the reactant concentration decreases. The size of the reactor is usually an investment decision.

However, excessive exposure to reaction conditions often reduces yield by formation of undesirable by products.

$$A + B \rightarrow C + D ------ E \text{ undesirable}$$

For the purposes of this course, the equation clearly states that to obtain a high conversion takes exponentially more time.

$$kt = \frac{x}{a(a-x)}$$

k = reaction rate constant
t = lapsed time
x = amount reacted
a = original concentration

Agitation (Mixing)

As stated many times, the key to effective mass and energy transfer is good contact between molecules. This is a key principle to good chemical reaction results. Contact can be enhanced by mechanical agitation (stirred), external circulation, shaking and etc. The purpose is to maintain uniform conditions of concentration and temperature (good mixing) throughout the entire reactor volume.

PTEC 2431   Industrial Processes
SECTION 12 Chemical Reactions   T.D. Felder

Catalyst (Figure 3)

A catalyst is a chemical that participates in a chemical reaction but is not chemically changed. Not all chemicals to be reacted do so by simply being brought together. From the Arrhenius equation, it is clear that a minimum amount of energy is required to get the reaction started. In some cases, raising the temperature of the reactants is all that is needed. However, the required temperature increase may be too high and lead to undesirable side reactions, phase changes, thermal degradation or other conditions that reduce yield or would require impractical operating conditions.

Materials called catalysts have been developed to solve the problem. A reaction can be started at a lower temperature, product mix controlled, reaction rate moderated and conversion increased by careful selection of a catalyst.

Some catalysts form intermediate products which react further to final products, and the catalyst is released to assist in another reaction.

Catalysts can be a liquid or a fixed bed (solid). They can be regenerated in place, separated for reuse, disposal of, or left in the product. The point is that they are not incorporated into the chemical structure of the product. Catalysts activity can be lost (deactivated) in numerous ways and must be regenerated or replaced.

Initiator

Special chemicals with catalytic properties are called initiators. This material starts reactions that otherwise would be impractical at reasonable conditions. A peroxide to initiate olefin polymerization is a typical example.

Inhibitor

Some materials have the ability to slow reactions down or to prevent undesirable reaction rates from taking place. They are called inhibitors. They also can be very useful in storage and shipping of unstable chemicals by preventing degradation

III. HOW THEY WORK

Batch Reactors (Figure 1)

Batch Reactors are mostly used in petrochemical plants to treat large streams that contain low concentrations of reactants. They are useful to obtain high conversion and produce products in low amounts per unit volume.

PTEC 2431   Industrial Processes

SECTION 12: Chemical Reactions   T.D. Felder

Many plants sample water and other wastes before disposal to insure regulatory conformity. The water may need chemical treatment such as pH adjustment or removal of insoluble impurites by decantation or skimming. Since the solutions are very dilute, temperatures low and mixing poor, the reaction rates are low. The long hold-up time provided by batch reactors is ideal for these conditions.

The large tanks used in water treatment provide the long hold-up time necessary to complete slow chemical reactions and allow settling of non-crystalline solids. In principle, they function much like batch reactors. The soda-lime process to soften water is a typical example. The long settling time for the solids supplied by long holdup time is a must for effective refrigeration.

Tubular Reactors (Figure 2)

Tubular reactors provide high velocity (good mixing). The hold-up time is short and heat transfer rate is high. The reactants concentration decreases with time so the reaction rate slows and the rate of temperature rise decreases.

Tubular reactors are used in several important, polymerization reactions. Due to the low holdup (volume) or reactants, some variation in reactor conditions can be safely tolerated.

When products are drawn from the main reactor, the reaction is not complete. That is because conversion is not 100%. The reactants and product concentrations are at reactor conditions in the make. If fed to another reactor, with no new reactants added, the reaction can approach completion. This process is called clean-up or a secondary reaction. A shell and tube heat exchanger may be used as the reactor for temperature control since additional cooling may be necessary.

In some reactors the temperature may be allowed to increase adiabatically to speed up the reaction to conclusion.

Kill Reactor

In some reactions, it is best to stop or kill the reaction as the products leave the main reactor. At times, this can be accomplished by a sharp reduction in the temperature. In catalytic reactions, it may be necessary to add a chemical to kill the reaction by deactivating the catalyst.

Adiabatic Reactor Figure 5)

The adiabatic reactor is one of the most commonly used in the petrochemical plants. With no heat removal required, a major process step and the related equipment are eliminated. The reaction temperature is controlled through this series of events.

PTEC 2431   Industrial Processes

SECTION 12: Chemical Reactions   T.D. Felder

1. The amount of material reacted is fixed by the reaction rate and hold-up time.

   Material reacted, lbs = reaction rate (lb\HR) x Hold-up time (HRS)

   $$\text{Hold-up Time} = \frac{\text{Reactor Volume (FT}^3\text{)}}{\text{Feed Rate (FT}^3\text{\textbackslash HR)}}$$

2. The heat released is now fixed.

   ΔH_R = ΔH_F reactants - ΔH_F products
   Heat of reaction   = Heats of formation of reactants - heats of formation of products

3. That fixes the ΔT (Temperature rise)

   $$\Delta T = \frac{\Delta H_R}{(M)(C_P)}$$

   C_P = BTU\lb\ °C
   M = reactor feed, lbs\HR
   ΔH_R = Heat of reaction, BTU\HR

4. Now the reactor temperature can be controlled by the feed temperature.

   $$T_R = T_F + \Delta T$$

All the above are fairly straight forward concepts so where does the heat of reaction go?

The heat sink is usually a solvent that dissolves both the reactants and products. The solvent must be inert and not take part in the reactions. The heat of reaction goes to heat the reactor make. The difference between feed temperature and make temperature is caused by the heat of reaction.

One necessary condition is easy separation of products from the solvent for recycle.

Good mixing is critical for mass and energy transfer and is usually accomplished through agitation.

Stirred Reactors With External Heat Removal Capability (Figure 4)

The reactor with external heat removal capability is another frequently used process in petrochemical plants. Its primary attraction is the higher concentration of reactants allowed in the feed to be converted to products which increase production rate per unit of reactor volume. The heat is removed continuously, usually in a circulating stream, through a cooler. One drawback is the increased possibility of a run away reaction if heat removal capability is interrupted.

PTEC 2431   Industrial Processes
SECTION 12: Chemical Reactions   T.D. Felder

Heat removal is accomplished in a number of other ways including a jacket on the reactor, coils inside the reactor and an external exchanger through which the reactor content is circulated. One special case is the use of a reflux condenser.

The reactor temperature is controlled through this series of events.

1. The amount of material reacted is fixed by the reaction rate and hold-up time.
   Material reacted, lbs = reaction rate, lb\HR x Hold-Up time, HRS.

   Hold-Up Time = $\dfrac{\text{Reactor Volume (FT}^3\text{)}}{\text{Feed Rate (FT}^3\text{\textbackslash HR)}}$

2. The heat released is now fixed.

   ΔH_R = ΔH_F reactants - ΔH_F products
   Heat of reaction   Heats of formation of reactants - heat of formation or products

3. At this point, the external heat removal capability comes into play. It would typically be a shell and tube heat exchanger in a circulating loop.

$\Delta T_R = (M)(C_P)(\Delta T)$ + $(M)(C_P)(\Delta T)$

M = Feed (lb\HR)	M = Circulation rate lb\HR
C_P = Heat capability BTU\lb.\°C	C_P = Heat capability BTU\lb.\°C
ΔT = Reactor temperature - feed temperature	ΔT = Temperature of process into cooler - temperature of process out of cooler

IV. TERMS AND SYMBOLS

<u>Reactor</u> is a piece of equipment designed to maintain the conditions necessary to carry out a chemical reaction.

<u>Hold-up time</u> is the period of time that the reactants are in contact with the products. It is simply the reactor volume divided by the feed rate in a continuous reaction process.

<u>Concentration</u> is the weight fraction or weight % or the mol fraction or mole % of any component in the system.

PTEC 2431 Industrial Processes
SECTION 12: Chemical Reactions
T.D. Felder

<u>Reaction rate</u> is the amount of product formed or the amount of reactant consumed at a fixed point in time.

<u>Adiabatic</u> describes a reaction system to which no heat is added or removed during the life of the reaction. The heat of reaction is usually taken up as a temperature increase (ΔT) in the products.

<u>Temperature rise or fall</u> is the increase or decrease in temperature due to the release or absorption of heat energy produced by the reaction.

<u>Heat of Reaction</u> is the energy released or absorbed as reactants form products. It is calculated as the difference in the algebraic sum of the heats or formation of the reactants and products.

<u>Conversion</u> is the lb.-moles of reactant changed to its atomic equivalent as product, by-product and etc.

<u>Yield</u> is the lb.-moles of reactant changed to its atomic equivalent in product.

<u>Yield Loss</u> is the lb-moles of reactant changed to its atomic equivalent in undesirable products.

<u>Exothermic</u> means that heat is released by the reaction.

<u>Endothermic</u> means that heat is absorbed by the reaction.

<u>Cracking</u> is thermal breakdown of large molecules into smaller molecules or elements.

<u>Reformer</u> is a reactor that recombines a mixture of molecules or elements into more useful products.

<u>Residence Time - Hold-Up time - Contact time</u> all mean the time that the reactants and products are in contact for reaction to occur.

<u>Activation Energy</u> is the energy required to start and maintain a reaction. Heat energy is usually used.

<u>Turndown</u> is the amount the production rate can be reduced and produce acceptable products.

<u>Value In Use</u> is a term that describes the increase in value of products made from refrigerants.

PTEC 2431 Industrial Processes

SECTION 12 Chemical Reactions   T.D. Felder

<u>Steady State</u> in a reaction means all significant variables are at a fixed value with time.

<u>Catalyst</u> is a material that moderates reaction conditions to a more desirable levelel but does not react itself.

<u>Initiator</u> is a material that starts a reaction at a lower activation energy. Typically, a peroxide to "initiate" a polymerzation reaction.

<u>Inhibitor</u> is a material that can slow a reaction down to a controlled rate. Also, an inhibitor can be useful to prevent undesirable reactions in the storage of unstable materials.

REACTANTS

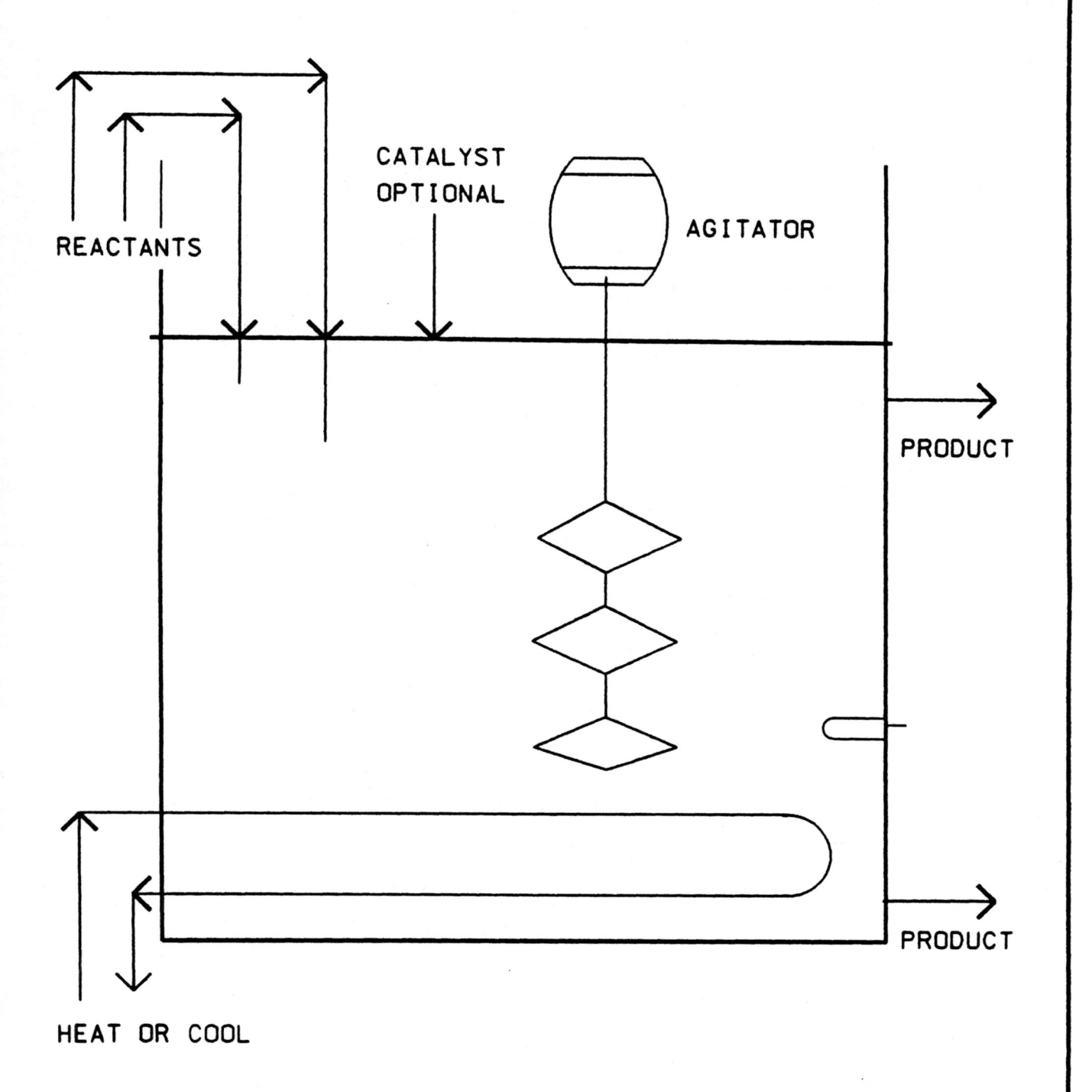

CATALYST OPTIONAL

AGITATOR

PRODUCT

PRODUCT

HEAT OR COOL

FIGURE
A <u>BATCH OR TANK REACTOR</u> OFFERS LONG HOLDUP TIME
AND HIGH CONVERSION FOR SLOW REACTIONS.

| THE VICTORIA COLLEGE | | 11/24/99 |

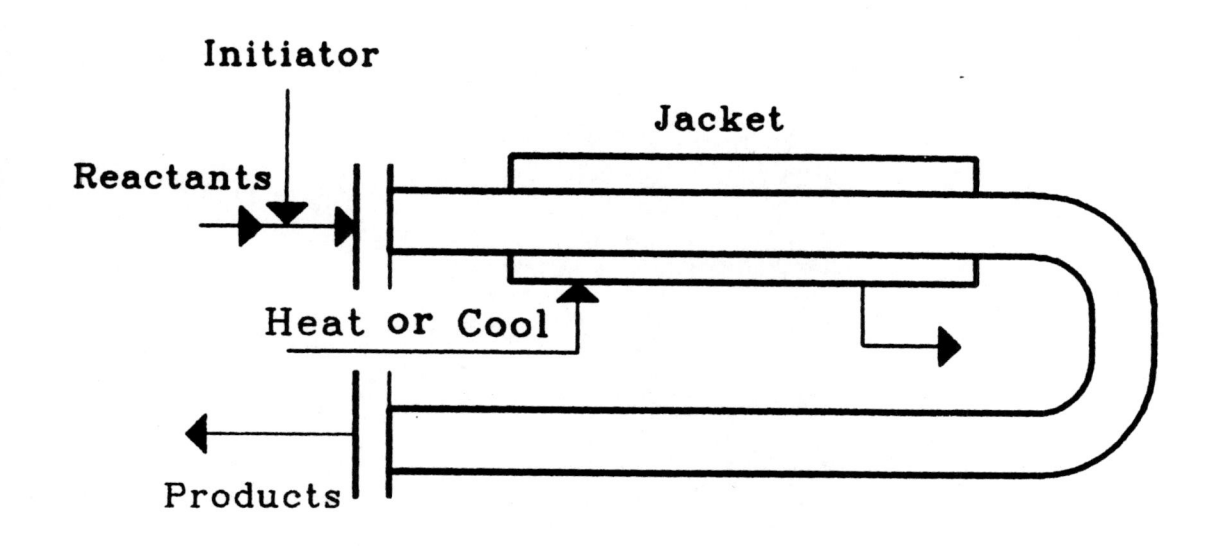

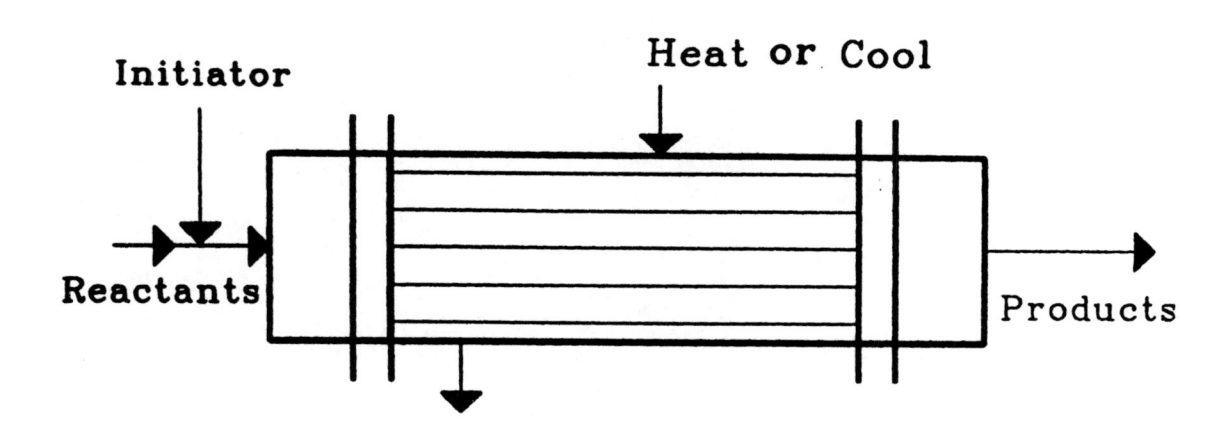

Figure TUBULAR REACTORS feature low holdup time and maximum mixing provided by high velocity.

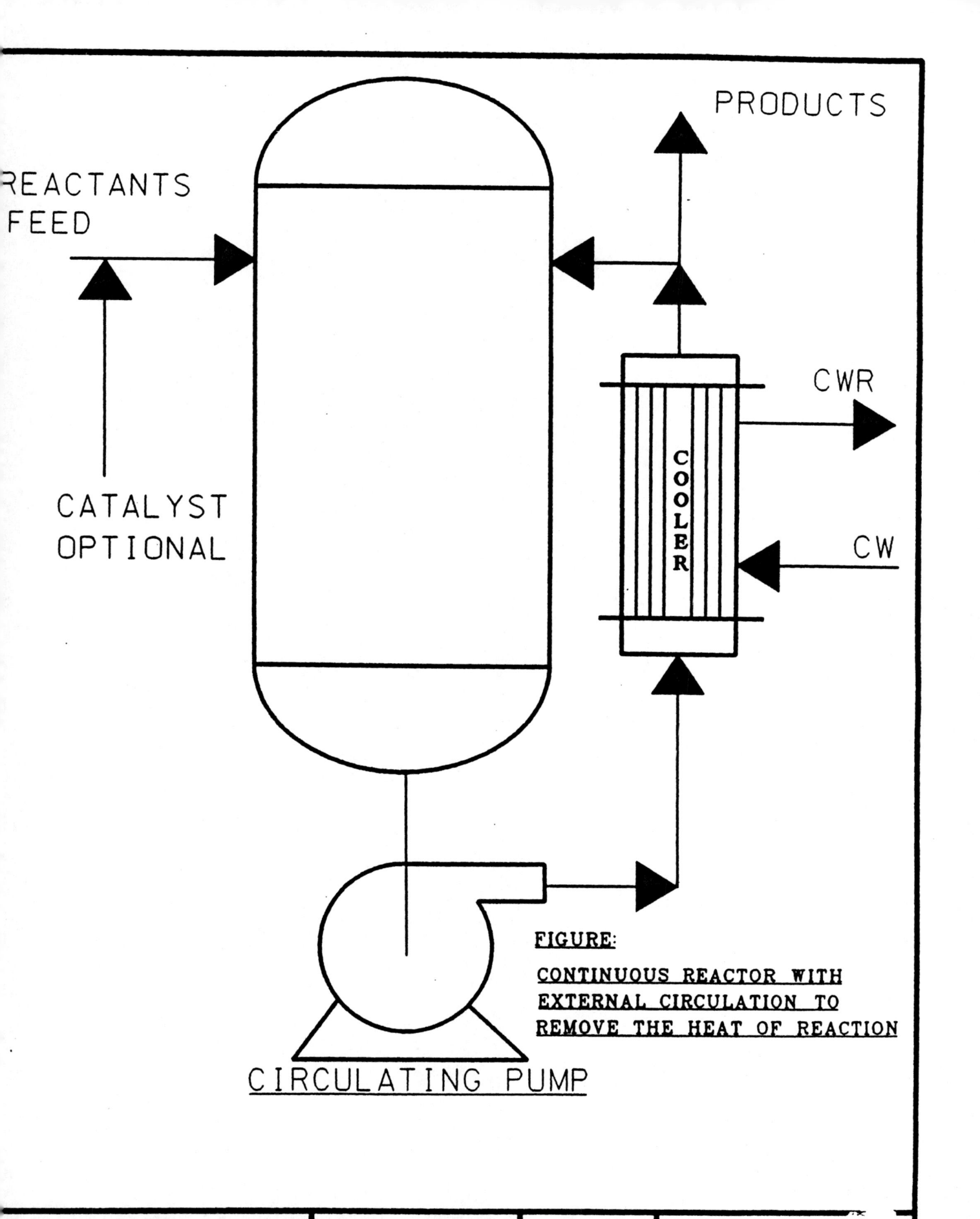

FIGURE:
CONTINUOUS REACTOR WITH EXTERNAL CIRCULATION TO REMOVE THE HEAT OF REACTION

| THE VICTORIA COLLEGE | FLOW DIAGRAM | 11-15-99 | |

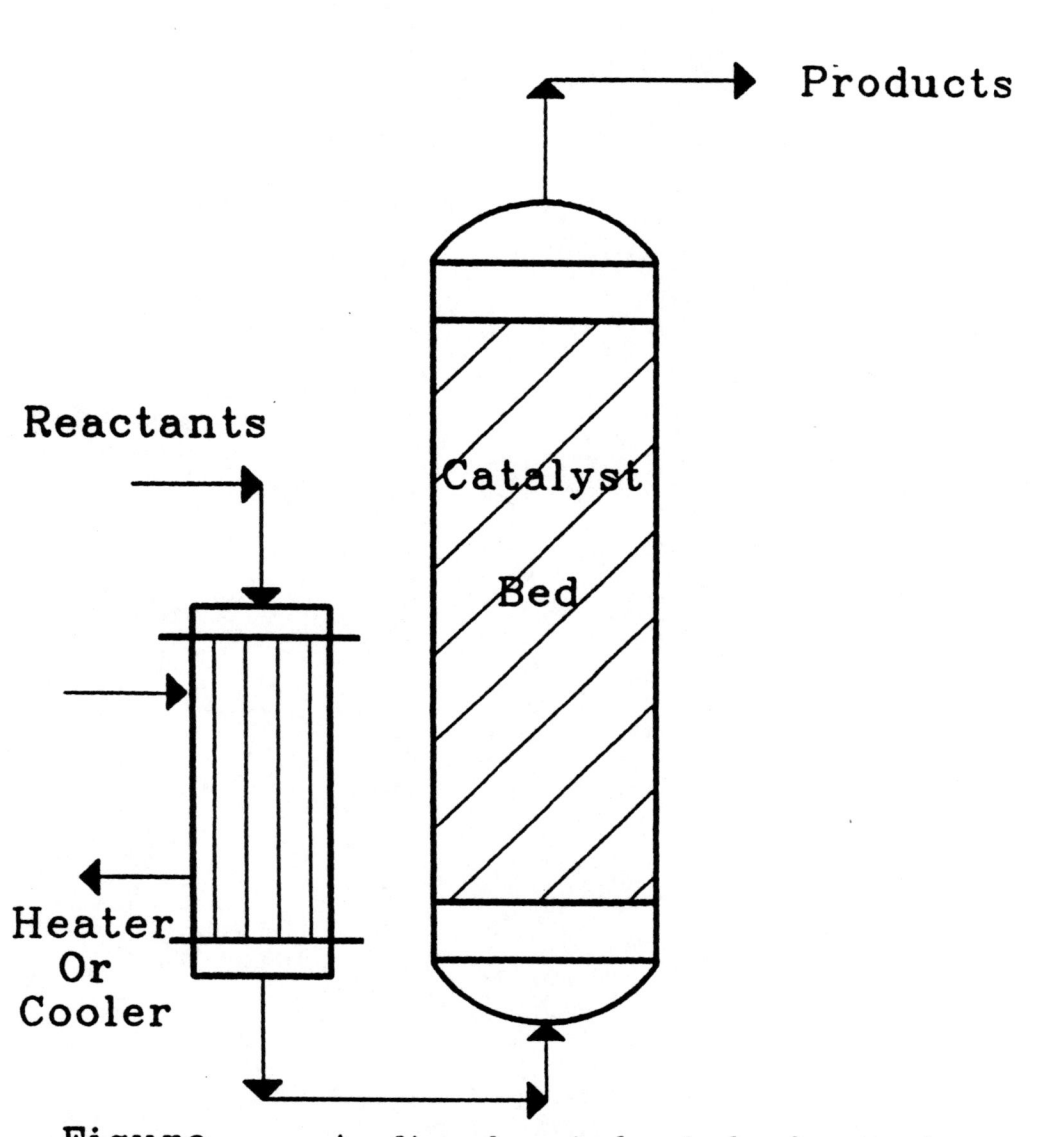

Figure A fixed catalyst bed reactor with a heat exchanger in the feed stream

| The Victoria College | | 11/12/99 | Flow Diagram |

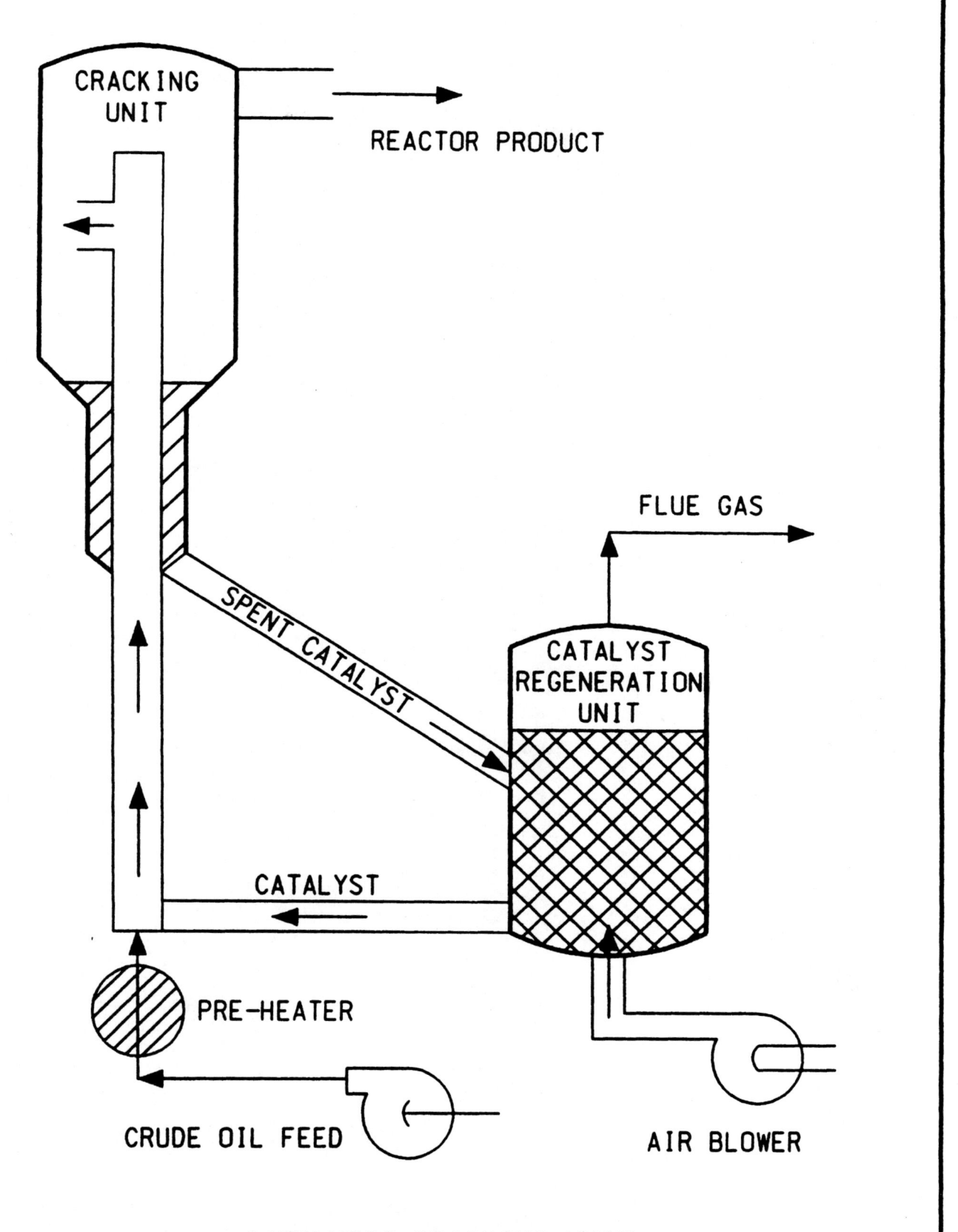

CATALYTIC CRACKING UNIT

STACK GAS

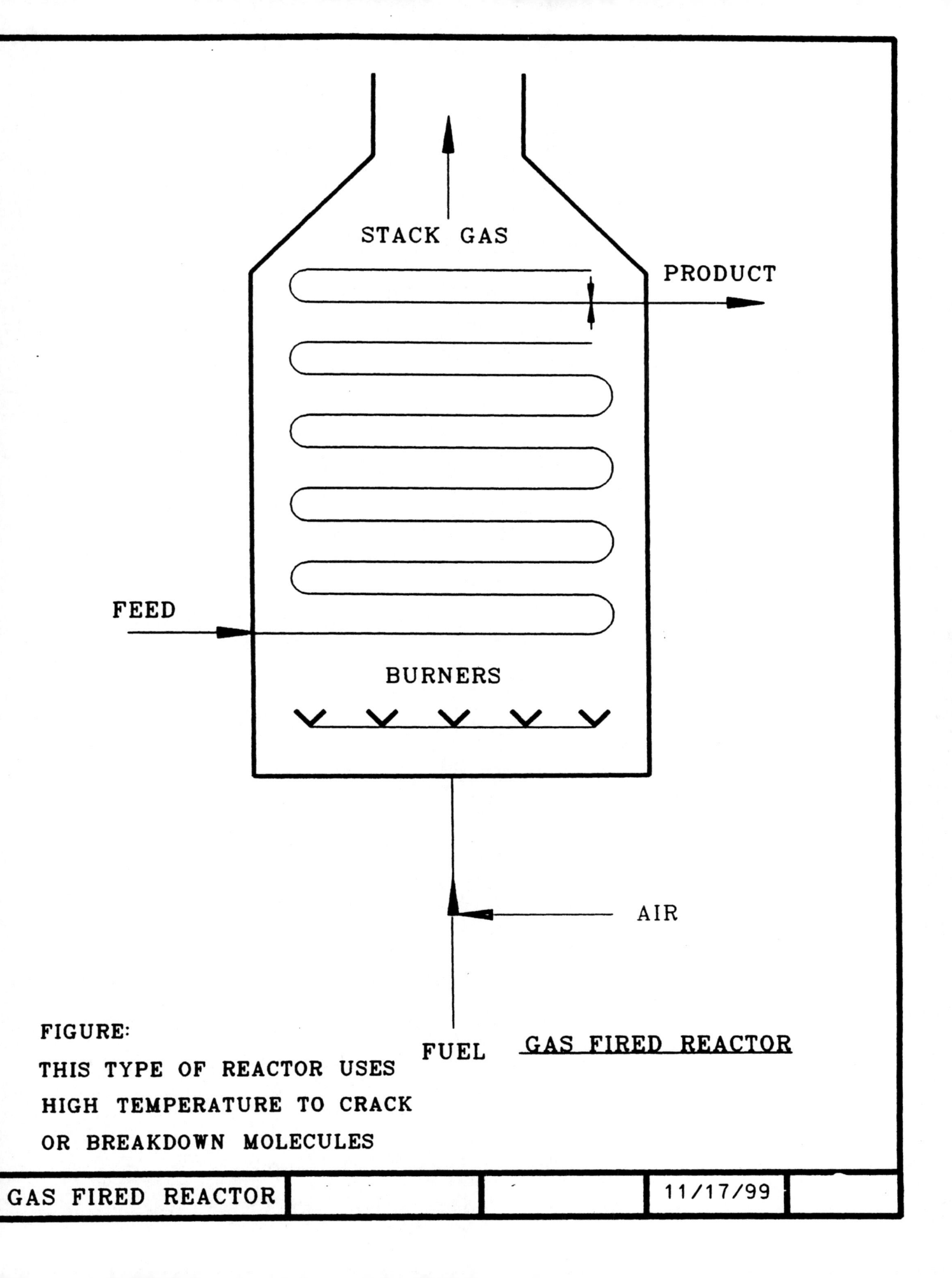

PRODUCT

FEED

BURNERS

AIR

FUEL   GAS FIRED REACTOR

FIGURE:
THIS TYPE OF REACTOR USES
HIGH TEMPERATURE TO CRACK
OR BREAKDOWN MOLECULES

| GAS FIRED REACTOR | | | 11/17/99 | |